PLANT SCIENCE RESEARCH AND PRACTICES

CORIANDER

DIETARY SOURCES, PROPERTIES AND HEALTH BENEFITS

PLANT SCIENCE RESEARCH AND PRACTICES

Additional books and e-books in this series can be found on Nova's website under the Series tab.

PLANT SCIENCE RESEARCH AND PRACTICES

CORIANDER

DIETARY SOURCES, PROPERTIES AND HEALTH BENEFITS

DEEPAK KUMAR SEMWAL
EDITOR

Copyright © 2019 by Nova Science Publishers, Inc.

All rights reserved. No part of this book may be reproduced, stored in a retrieval system or transmitted in any form or by any means: electronic, electrostatic, magnetic, tape, mechanical photocopying, recording or otherwise without the written permission of the Publisher.

We have partnered with Copyright Clearance Center to make it easy for you to obtain permissions to reuse content from this publication. Simply navigate to this publication's page on Nova's website and locate the "Get Permission" button below the title description. This button is linked directly to the title's permission page on copyright.com. Alternatively, you can visit copyright.com and search by title, ISBN, or ISSN.

For further questions about using the service on copyright.com, please contact:
Copyright Clearance Center
Phone: +1-(978) 750-8400   Fax: +1-(978) 750-4470   E-mail: info@copyright.com.

NOTICE TO THE READER

The Publisher has taken reasonable care in the preparation of this book, but makes no expressed or implied warranty of any kind and assumes no responsibility for any errors or omissions. No liability is assumed for incidental or consequential damages in connection with or arising out of information contained in this book. The Publisher shall not be liable for any special, consequential, or exemplary damages resulting, in whole or in part, from the readers' use of, or reliance upon, this material. Any parts of this book based on government reports are so indicated and copyright is claimed for those parts to the extent applicable to compilations of such works.

Independent verification should be sought for any data, advice or recommendations contained in this book. In addition, no responsibility is assumed by the Publisher for any injury and/or damage to persons or property arising from any methods, products, instructions, ideas or otherwise contained in this publication.

This publication is designed to provide accurate and authoritative information with regard to the subject matter covered herein. It is sold with the clear understanding that the Publisher is not engaged in rendering legal or any other professional services. If legal or any other expert assistance is required, the services of a competent person should be sought. FROM A DECLARATION OF PARTICIPANTS JOINTLY ADOPTED BY A COMMITTEE OF THE AMERICAN BAR ASSOCIATION AND A COMMITTEE OF PUBLISHERS.

Additional color graphics may be available in the e-book version of this book.

Library of Congress Cataloging-in-Publication Data

ISBN: 978-1-53616-483-1
Library of Congress Control Number:2019951539

Published by Nova Science Publishers, Inc. † New York

CONTENTS

Preface		vii
Acknowledgments		xi
Acronyms		xiii
Chapter 1	Coriander: A Nutritional and Functional Food *Siddharth Priyadarshi, R. Chetana and Madeneni Madhava Naidu*	1
Chapter 2	Dried Coriander: Processing and Properties *Raquel P. F. Guiné, Sofia G. Florença and Maria João Barroca*	37
Chapter 3	Antioxidant and Antimicrobial Activities of Coriander (*Coriandrum sativum*) *Miroslava Kačániová and Eva Ivanišová*	63
Chapter 4	*Coriandrum sativum:* A Plant of Health Benefits and Biotechnological Applications for Improvement *Muzamil Ali, A. Mujib, Nadia Zafar and Basit Gulzar*	95

Chapter 5	Coriander Seed as a Source of Biologically Active Compounds: Essential Oil, Chemical Composition and Biological Activity *Saša Đurović and Stevan Blagojević*	139
Chapter 6	Polyphenolic Compounds of Coriander Plant for Human Health and Diseases *Arjun Pandian, Raju Ramasubbu, Kaliyaperumal Ashokkumar, Ruchi Badoni Semwal, Sudharshan Sekar and Samiraj Ramesh*	183
Chapter 7	Essential Oil of Coriander: A Source of Antimicrobial Agent *Sonali Aswal, Ankit Kumar, Ashutosh Chauhan and Deepak Kumar Semwal*	207

About the Editor 235

Index 237

Related Nova Publications 247

PREFACE

The present reference book is based on the nutritional and medicinal importance of 'coriander' which is one of the most used herbs for culinary and medical purposes. Coriander is scientifically known as *Coriandrum sativum* L. and is a member of the Umbelliferae family. This herb is one of the oldest traditional medicines that have been used for more than 3000 years for curing a variety of ailments including indigestion, worm infections, rheumatism, loss of appetite, convulsion, insomnia, anxiety and joints pain. As a folk medicine, this herb is known for its carminative, spasmolytic, digestive and galactagogue properties.

Coriander is used as one of the main ingredients in the preparation of many Ayurvedic formulations. The plant is grown throughout the world mainly for its seeds, leaves and essential oil. It is used as a flavouring agent for sweets, beverages and baked products. Coriander is found to be rich in volatile oil with linalool is as a major constituent together with α-pinene and terpinene. Different parts of this plant contain monoterpenes, α-pinene, limonene, γ-terpinene, p-cymene, borneol, citronellol, camphor, geraniol, coriandrin, dihydrocoriandrin, coriandrons A-E, coumarins, phthalides, flavonoids and other phenolic acids. This is an edible herb and there are no toxic reports available on this plant, in other words, the herb is non-toxic to humans and animals.

Coriander seeds and essential oil have been extensively investigated for their chemical composition and biological activities. Various parts of this plant possess antidiabetic, laxative, diuretic, tonic, hypolipidemic and anticancer effects. It has been reported to have strong fungicidal and bactericidal properties. Its essential oil has been exhibited potent antifungal activity even at very low concentration. Although this herb has good commercial importance, however, physical properties, chemical composition and bioactivity can affect its commercial value.

Essential oils are volatile in nature and comprised of various small molecules such as terpenes, hydrocarbons and alcohols. These are soluble in most of the organic solvents including ether, chloroform and ethanol. These can be extracted via steam distillation or expression method. Steam distillation or hydrodistillation is a conventional method used to extract most of the oils whereas expression method is used for selective cases such as extraction of citrus essential oils by cold pressing.

Till the date, approximately 100 essential oils are known to use in medicine and cosmetic products. In addition to the therapeutic benefits, these are also used as flavours to food and beverages, perfumes, soap, shampoos and massage oils. Some of the most common health problems like stress, anxiety, depression, headaches, migraines, sleep, insomnia, inflammation, antifungal and antimicrobial infections can be treated with essential oil in many ways. Due to the antimicrobial and disinfectant properties, the essential oils are used in dentistry adhesives, floor cleaners and insect repellents.

This reference book entitled "Coriander: Dietary Sources, Properties and Health Benefits" is comprised of seven chapters, contributed by different authors, covering whole information about this wonderful herb. Its occurrence, taxonomy, traditional uses, phytochemistry and pharmacological activities are well described with supporting references.

The information provided in the book will be helpful for students, academicians and scientists working in the field of plant sciences, natural products and other relevant areas. Its cultivation, processing, commercial uses, mainly of its essential oil, are described in a simple language by keeping in mind a common man and agriculturist.

Deepak Kumar Semwal, PhD
Editor

ACKNOWLEDGMENTS

Being an Editor, I thank authors for their valuable contribution and kind support during the editing and composing process. They have been well supportive and responsive during the whole process.

The contents of the book have been extensively reviewed by different subject experts, and based on their opinions, the quality of the work has been improved. The contribution of the reviewers is much appreciated as without their support, this form of the book was never possible. Hence, I must pay my sincere thanks to the reviewers for their invaluable time to accept my request for reviewing the book contents.

I am thankful to Nova Science Publishers Inc. for inviting me to edit this book on coriander, one of the most commercially important plants in the world.

I wish to thank the reviewers/subject experts for sparing their precious time to accept the review invitation. Indeed, their valuable opinions/ comments have been found worthy to improve the contents of this book. I would like to mention the name of the reviewers who contributed to this book by giving their expertise.

Prof. R. S. Dahiya, Maharishi Markandeshwar University, Ambala, India.

Prof. João Carlos Gonçalves, Agrarian School of Polytechnic Institute of Viseu, Portugal.

Prof. Swapnil Singhai, Uttarakhand Ayurved University, Dehradun, India.

Prof. Maria João Dias, Agrarian School of Polytechnic Institute of Coimbra, Portugal.

Dr. Debabrata Sircar, Indian Institute of Technology, Roorkee, India.

Dr. Skelwa Cosa, University of Pretoria, South Africa.

Dr. P. Arjun, PRIST University, Tamil Nadu, India.

Dr. Ruchi B. Semwal, Pt. L.M.S. Govt. PG College, Rishikesh, India.

Dr. Ashutosh Chauhan, Uttarakhand Ayurved University, Dehradun, India.

Dr. Aijaz Ahmad, University of the Witwatersrand, Johannesburg, South Africa.

ACRONYMS

ABTS	2,2'-Azinobis-(3-ethylbenzothiazoline-6-sulfonate)
ALP	Alkaline phosphatase
ANN	Artificial neural network
BC	Before Christ
BHA	Butylated hydroxyl anisole
BHT	Butylated hydroxyl toluene
BuOH	Butanol
BW	Body weight
CCl_4	Carbon tetrachloride
CE	Catechin equivalents
DCM	Dichloromethan
DEE	Diethyl ether
DM	Dry mass
DMH	1,2-Dimethylhydrazine
DNA	Deoxyribonucleic acid
DPPH	2,2-Diphenyl-1-picrylhydrazyl
DW	Dry weight
EA	Ethyl acetate
EO	Essential oil
EtOH	Ethanol
FRAP	Ferric Reducing Antioxidant Power Assay

FW	Fresh weight
GAE	Gallic acid equivalent
H_2O	Water
HDL	High-density lipoprotein
HRMS	High resolution mass spectrometry
LC_{50}	Lethal concentration required to kill 50% of the population
LDL	Low-density lipoprotein
LPS	Lipopolysaccharide
MAPK	Mitogen-activated protein kinase
MBC	Minimum bactericidal concentration
Me	Methyl
MeOH	Methanol
MFC	Minimum fungicidal concentration
MIC	Minimum inhibitory concentration
NMR	Nuclear magnetic resonance
ORAC	Oxygen radical absorbance capacity
ROS	Reactive oxygen species
RSM	Response surface methodology
scCO2	Supercritical carbon dioxide
SGOT	Serum glutamic oxaloacetic transaminase
SGPT	Serum glutamic pyruvic transaminase
SWE	Subcritical water extraction
TE	Trolox equivalent
UAE	Ultrasound-assisted extraction
UV	Ultraviolet
W/V	Weight by volume
W/W	Weight by weight

In: Coriander
Editor: Deepak Kumar Semwal

ISBN: 978-1-53616-483-1
© 2019 Nova Science Publishers, Inc.

Chapter 1

CORIANDER: A NUTRITIONAL AND FUNCTIONAL FOOD

Siddharth Priyadarshi[1,2], R. Chetana[3] and Madeneni Madhava Naidu[1,2,]*

[1]Academy of Scientific and Innovative Research (AcSIR), CSIR-CFTRI Campus, Mysuru, Karnataka, India
[2]Department of Spices and Flavour Sciences, CSIR-CFTRI, Mysuru, Karnataka, India
[3]Department of Traditional Food and Sensory Science, CSIR-CFTRI, Mysuru, Karnataka, India

ABSTRACT

This chapter deals with the literature survey relating to the properties and uses of coriander foliage as well as seeds. The history, cultivation, plant characteristics, botanical classification, nutritional profile, phytochemical content, proximate composition, applications, medicinal and pharmacological properties of coriander leaf and seeds are extensively reviewed. One section describes the green coriander foliage as a spice and

---

*Corresponding Author's Email: mmnaidu@cftri.res.in.

vegetable with particular emphasis on its culinary applications. The bioactive compounds in coriander (seeds and foliage) and their medicinal and pharmacological activities have been reviewed. The culinary applications, nutritional, nutraceuticals and antioxidant properties are well discussed. The information suggests that the coriander leaf and seeds provide important micro and macro nutrients and are good source of natural bioactive compounds having varied health benefits.

Keywords: coriander, nutritional profile, nutraceuticals, applications, pharmacological properties

INTRODUCTION

Coriander (*Coriandrum sativum* L.) has a long history of use for culinary and medicinal purposes. It is mentioned in Sanskrit literature as far back as 5000 B.C. and in the Greek Eber Papyrus as early as 1550 B.C. (Uhl 2000). It was used in traditional Greek medicine by Hippocrates (ca. 460–377 B.C.). The Egyptians called this herb "the spice of happiness" probably because it was considered to be an aphrodisiac (Grieve 1971). Coriander is an important spice crop and inhabits a critical position in flavouring substances. It is a delicately branched annual herb that belongs to the family *Apiaceae*. The whole plant and mainly the unripe fruit when rubbed give characteristic aroma and strong odour, hence the name coriander (Pathak et al. 2011). Coriander has a pleasant aromatic fragrance. It is cultivated as a domestic plant (Carrubba 2002).

Coriander is a tropical crop. It requires dry frost-free cold climate at flowering and seedling stage for excellent quality and higher yield. The ideal temperature for germination and early development of coriander is 20-25°C (Rajeshwari and Andallu 2011). Coriander plants can tolerate pH levels ranging from 4.8 to 8.2. During dry weather conditions, it requires total irrigation and sunlight. The coriander seeds should be sown during spring after the last frosting. It requires 10 to 20 days for the seedling to appear. The high nitrogen content during excessive fertilization hinders the seed maturation as well as reduces the flavour. Coriander is cultivated in the Asia

Minor, Caucasus, the Mediterranean region and Southern Europe. In recent years, principal commercial coriander producers included members of the former Soviet Union, Hungary, Poland, Romania, Russia, Czech Republic, Slovakia, Morocco, Ukraine, Canada, India, Pakistan, Iran, Turkey, Guatemala, Mexico, and Argentina (Diederichsen 1996).. The seeds have diuretic and carminative property and are also used in the preparation of many traditional medicines to cure bed cold, nausea, seasonal fever and stomach disorders (Purseglove et al. 1981).

Coriander provides an aromatic flavour that is indicative of both citrus peel and sage. It originated in the Mediterranean regions and is cultivated mainly in tropical areas (Diederichsen 1996). Global production of coriander seed was estimated to be 6.25 Lakhs metric ton (MT) in the year 2015-16. India is the most significant producer of coriander seed in the world. India produced around 5 Lakhs MT/year, which contributed to 80% of the total coriander seed production in the world. Spice Board of India reported that India exported about 40,100 MT with a total value of Rs. 234 Crore in the year 2015-16. The main coriander growing states in India are Andhra Pradesh, Madhya Pradesh, Rajasthan and Tamil Nadu. Other coriander producing states in India are Uttar Pradesh, Himachal Pradesh, Gujarat, Uttaranchal, Bihar, West Bengal, Jharkhand, Chhattisgarh, Karnataka and Orissa. The major markets in India are Madhya Pradesh and Rajasthan (Kota, Ramganj Mandi and Baran) (Priyadarshi and Borse 2014).

The intraspecific classification of coriander depends on the description of the diversity of the species. De Candolle in the year 1830 gave the first intraspecific classification of coriander; thus coriander bearing small fruits was described as *Coriandrum sativum* L. var. *microcarpum DC* after his name. Coriander, having large fruits, is found not only as a crop in Europe but also as a weed (Diederichsen 1996). Alefeld in the year 1866 described coriander with large fruits as *Coriandrum sativum* L. var. *vulgare alef*. According to intraspecific classification, if the weight of 1000 seeds is less than 10 g, then this coriander plant belongs to *Coriandrum sativum* L. *microcarpum DC*, whereas; plants bearing larger seeds with more weight are considered as *Coriandrum sativum* L. *vulgare alef* (Diederichsen 1996). International Code of Botanical Nomenclature suggested that the name of

Linnaeus is autonomous (Greuter 1994), hence, *Coriandrum sativum* L. var. *sativum* should be used in place of the *Coriandrum sativum* L. *vulgare alef*. Diederichsen (1996) stated that the diameter and weight of 1000 seeds bears a very high significant correlation (r^2=0.92). Consequently, an infraspecific classification identified two groups of coriander as:

- *Coriandrum sativum* L. var. *microcarpum* DC: diameter of seeds less than or equal to 3 mm and weight of 1000 seeds less than 10 g.
- *Coriandrum sativum* L. var. *sativum*: diameter of seeds more than 3 mm and weight of 1000 seeds more than 10 g.

The genus *Coriandrum* L. has two species; large fruited *C. sativum* L. var. *vulgare alef* and small-fruited *C. sativum* L. var. *microcarpum* DC (Rajeshwari and Andallu 2011). *Coriandrum sativum* L. var. *vulgare alef* is represented as Russian coriander which bears large fruits rich in the volatile oil. While *Coriandrum sativum* L. var. *microcarpum* DC carries smaller fruits having low volatile oil and are found in India, Morocco and other Asian countries. It reaches a height of 2-3 feet (0.6-0.9 m) with a spread of 1-2 feet (0.3-0.6 m) (Purseglove et al. 1981; Singh and Ramanujam 1973).

Although now popular and fashionable herb coriander is used to flavour as well as a garnish, it is mainly cultivated for its seeds (Wangensteen et al. 2004). It has been reported to have some possible health attributes, particularly relating to the gastrointestinal tract, but also as a possible diabetic remedy (Gray and Flatt 1999; Al-Mofleh et al. 2006). However, most of these remedies and most studies utilise seeds, rather than leaves. Variation in numerous traits makes coriander appropriate for various uses.

BOTANICAL DESCRIPTION

Coriandrum sativum L. belongs to the family *Apiaceae* with botanical classification shown in Table 1. Coriander is a small herb bearing several sub-branches and branches. The new leaves are oval, but aerial leaves are elongated in shape. The coriander flowers are white, while fruits are round shaped having brinjal like shades (Sharma and Sharma 2012).

Table 1. Botanical classification of coriander

Division	Angiospermae
Series	Calyciflorae
Class	Dicotyledonae
Sub-class	Polypetalae
Order	Umbellale (Apiales)
Family	*Umbelliferae (Apiaceae)*
Genus	*Coriandrum*
Species	*Coriandrum sativum*

Coriander is mostly consumed in all the parts of India and coriander is known by different names in various states of India. The various names in different states of India are - Telugu: Dhaniyalu; Tamil: Kothamali; Punjabi: Dhania; Oriya: Dhania; Marathi: Dhana, Kothimber; Malayalam: Kothumpkalari bija, Kothumpalari; Kashmiri: Daaniwal, Kothambalari; Kannada: Kothambri, Kothmiri bija; Gujarati: Kothmiri, Konphir, Libdhane and Bengali: Dhane, Dhania (Priyadarshi and Borse 2014). The green coriander plant is known as 'Kinza' in Russia, 'Chinese parsley' in England, 'Cilantro' in Spain and America, 'Persil arabe' in France and 'venshivu' and 'ounshavu' in Egypt (Diederichsen 1996).

NUTRITIONAL PROFILE

The nutritional profile for coriander leaves and seeds are shown in Table 2. Coriander leaves and seeds are free from cholesterol, Vitamin B-12 and Vitamin D (USDA, 2013). The most abundant mineral in coriander seed is potassium (1267 mg/100 g) followed by calcium (709 mg/100 g), phosphorus (409 mg/100 g), magnesium (330 mg /100 g), sodium (35 mg/100 g) and zinc (4.70 mg/100 g). Coriander leaves contain a high amount of vitamin C (566.7 mg/100 g) (Table 2). Coriander plant is a rich source of linalool in essential oil and petroselinic acid isolated from its aerial parts and seeds (Sahib 2012). The distillation residues contain a high amount of proteins and fats thus making it suitable for animal feed.

Table 2. Nutritional composition of coriander leaf and seed

Nutrients	Amount (per 100g)	
	Coriander leaf	Coriander seed
Calcium	1246 mg	709 mg
Carbohydrates	52.10 mg	54.99 mg
Cholesterol	0.00 mg	0 mg
Energy	279 kcal	298 kcal
Fatty acids, total monosaturated	2.232 mg	13.580 mg
Fatty acids, total polyunsaturated	0.328 mg	1.750 mg
Fatty acids, total saturated	0.115 mg	0.990 mg
Fiber, total dietary	10.40 mg	41.9 mg
Iron	42.46 mg	16.32 mg
Magnesium	694 mg	330 mg
Niacin	10.707 mg	2.130 mg
Phosphorous	481 mg	409 mg
Potassium	4466 mg	1267 mg
Protein	21.93 mg	12.37 mg
Riboflavin	1.500 mg	0.290 mg
Sodium	211 mg	35 mg
Thiamine	1.252 mg	0.239 mg
Total fat	4.78 mg	17.77 mg
Vitamin A, IU	5850 IU	0 IU
Vitamin A, RAE	293 µg	0.00 µg
Vitamin B-12	0.00 IU	0.00 IU
Vitamin C	566.7 mg	21.0 mg
Vitamin D	0 IU	0 IU
Water	7.30%	8.86%
Zinc	4.72 mg	4.70 mg

PHYTOCHEMICAL PROFILE

Coriander seeds contain 1.8% essential oil, containing major active chemical constituents and the volatile oil contains 65 to 70% of (+)-linalool (Verma et al. 2011). Coriander also contains high amounts of fatty acids ranging between 9.9 and 27.7% (Pathak 2011). It is reported that linolenic and linoleic acids are the major fatty acids present in coriander and the primary compounds present in coriander seed is presented in Table 3. These compounds are classified as aliphatic alcohols, aliphatic aldehydes, aliphatic hydrocarbons, monoterpene oxides and carbonyls, monoterpene alcohols, monoterpene esters, monoterpene hydrocarbons, *phenols,* sesquiterpenes and some miscellaneous compounds such as acetic acid and α-pdimethyl styrene (Parthasarathy et al. 2008). Coriander oil mainly contains vebriniol, jireniol and coriandrol (Rao 2012). The other minor components present in coriander are monoterpene hydrocarbons, heterocyclic compounds coriandrones A-E, dihyrocoriandrin, flavonoids, glazonoids, isocoumacinvizcoriandrin, phenolic acids, phthalides viz - neochidilide, sterols and Z-digustilide (Rao 2012).

The major monoterpene hydrocarbons present in coriander are borneol, citronellol, geraniol, geranyl acetate, limonene, xmphoe, α- pinene, β-pinene, γ-terpinene and ρ-lymene; whereas, major hetero-cyclic compounds in coriander include furan, pyrazine, pyridine, tetrahydrofuran derivatives and thiazole (Wallis 2005). The various phytonutrients present in coriander are endowed with a wide range of pharmacological benefits such as stomachic, stimulant, lipolytic, insecticidal, hypolipidemic, hypoglycemic, hepatoprotective, fungicidal, diuretic, diaphoretic, cytotoxic, carminative, antispasmodic, antimutagenic, antibacterial, and aflatoxin protective potential (Benjumea et al. 2005; Pandey et al. 2011).

Table 3. Classification of volatile compounds in coriander seed

S.No.	Compounds	Constituents	Reference
1.	Aliphatic alcohols	Decanol, dodecanol	Parthasarathy et al. 2008; Ramezani et al. 2009
2.	Aliphatic aldehydes	Octanal, nonanal, decanal, undecanal, dodecanal, tridecanal, tetradecanal, 3-octenal, 2-decenal, 5-decenal, 8-methyl-2-nonenal, 8-methyl-5-nonenal, 6-undecenal, 2-dodecenal, 7-dodecenal, 2-tridecenal, 8-tridecenal, 9-tetradecenal, 10-pentadecenal, 3,6-undecadienal, 5,8- tridecadienal	
3.	Aliphatic hydrocarbons	Heptadecane, octadecane	
4.	Monoterpene oxides and carbonyls	Camphor, 1,8- cineole, linalol oxide, carvone, geranial	
5.	Monoterpene alcohols	Borneol, citronellol, geraniol, linalool, nerol, α-terpineol, 4-terpinenol	
6.	Monoterpene esters	Bornyl acetate, geranyl acetate, linalyl acetate, α-terpinyl acetate	
7.	Monoterpene hydrocarbons	p-Cymene, camphene, Δ-3-carene, limonene (dipentene), myrcene, cis- and trans-ocimene, α-phellandrene, β-phellandrene, α-pinene, β-pinene, sabinene, α-terpinene, γ-terpinene, terpinolene, α-thujene	
8.	Phenols	Anethole, myristicin, thymol	
9.	Sesquiterpenes	β-Caryophyllene, caryophellene oxide, elemol, nerolidol	
10.	Miscellaneous compounds	Acetic acid, α-pdimethyl styrene	

USES OF CORIANDER

The two significant products obtained from coriander are fresh green foliage and ripe fruits. The leaves and mature fruits have a fresh and pleasant flavour. Mostly dried seeds and fresh leaves are used in cooking. Both are utilised for flavouring of foods, perfumes and cosmetics (Parthasarathy et al. 2008). The fresh plant (stem and leaves) is used as a herb for culinary purposes, while dried fruit is used as a spice. Coriander leaves are also known as cilantro, Chinese parsley, fresh coriander or coriander leaves. The foliage looks very fresh, firm, deep green without any brown or yellow spots. Many people like the strong smell of foliage, while some dislike its soapy taste. Seeds have a different taste compared to the citrus flavour of leaves. The leaves deteriorate quickly as soon as it is removed from the plant and tend to lose its taste and flavour when frozen or dried (Rajeshwari and Andallu 2011). Coriander foliage is utilised for its distinct colour, flavour and aroma. Leaves bear the citrus flavour while seed has different taste. Coriander is mostly utilised for the culinary purpose, as a condiment and garnish throughout the world (Anitescu et al. 1997; Bandoni et al. 1998). Both seeds and leaves are mainly used all over the world as volatile isolate form for flavouring sweets, tobacco products, beverages, baked goods and as a vital ingredient for curry powder. Fruits give essential oil ranging from approximately 0.5 to 2.5% which is used in the manufacture of soaps, perfumes and in flavours (Parthasarathy et al. 2008). The characteristic flavour in green herb is due to aldehydes present in the essential oil. These aldehydic compounds tend to decrease with ripening and are absent in fruits after complete maturation and drying (Diederichsen 1996). The American Midwest, Thailand, South America, Mexico, Malaysia, Iraq, Iran, Indonesia, India, China and Caucasus utilises coriander in the form of leaves (Prakash 1990).

Coriander is also used as a medicinal herb, to treat stomach disorders and as a digestive aid. Coriander is also utilised to treat diarrhoea and as a medication for oral infections (Chaudhry and Tariq 2006). Coriander seed is used as a natural food additive and is a rich source of natural antioxidants having antidiabetic activity (Gray and Flatt 1999; Eidi et al. 2009; Msaada et al. 2017). Coriander foliage has substantial health benefits as it is rich in beta-carotene, folic-acid, vitamin-A, vitamin-C, calcium, iron, magnesium, manganese, and potassium. It is best utilised while it is fresh as it holds its unique aromatic flavour and aroma. Coriander foliage is the most consumed leaves in diets of Malaysia, Singapore, Thailand and Indian cuisines. Foliage contains various bioactive compounds such as alkaloids, flavonoids and polyphenols which are responsible for its various biological activities. It is utilised as a home cure for treating fever, vomiting and stomach related problems (Ganesan et al. 2013).

Coriander leaves have high antioxidant activity as it contains a high amount of natural antioxidant vitamin-C and gives 30% recommended daily levels of ascorbic acid. Coriander leaves impart various health benefits comprising an antioxidant, antibiotic, anti-cancer, anti-inflammatory, antimicrobial properties, cholesterol-lowering ability, heavy metal detox, immune booster, improving eyesight, insomnia, lowering blood sugar, increased bone health, lowering the risk of kidney stones. It also reduces bad cholesterol (low-density lipoprotein) and raises good cholesterol (high-density lipoprotein) in the body. Coriander foliage has a wide range of benefits, making it a precious herb for mitigating numerous ailments (Rajeshwari and Andallu 2011).

Coriander leaves are enriched with quercetin, kaempferol and acacetin flavonoids. The phenolic acids identified were vanillic acid, ferulic acid (Trans and Cis form) and p-coumaric acid by paper and thin-layer chromatography (Nambiar et al. 2010). Foliage also contains 4-hydroxycoumarin apigenin, dicoumarin, diosmin, esculin, gallic acid, luteolin, tartaric acid and vicenin which are responsible for its antidepressant, antidiabetic, antimutagenic, and antioxidant activities (Kansal et al. 2011, Kansal et al. 2012). Maroufi et al. (2010) reported that the coriander foliage is also utilised for treatment of headache, abdominal discomforts, dry skin, eczema, erysipelas, inflammation and loss of appetite. Different parts of coriander have different uses. The traditional applications of the plant, which are based on the primary products, i.e., the fruits and the green herb, are two-fold: culinary and medicinal.

CORIANDER FOLIAGE AS A SPICE AND VEGETABLE

Coriandrum sativum L. is commonly known as dhania possessing various health benefits. The coriander oil functions as a crucial ingredient in curry mix and chiefly used around the world in the ground or volatile isolate form in flavouring beverages, bakery products, sweets, tobacco goods, etc. Coriander is a rich source of micronutrients and nutritional elements such as calcium, magnesium, sodium, potassium, dietary fibre and vitamins. Coriander leaves and seeds are used mostly in cooking. They also strengthen the stomach; reduce fever and lower cholesterol levels. The fruit of coriander plant comprises approximately 1% of volatile (essential) oil - an active element of this herb. This volatile oil is colourless or has a pale yellow hue with the fragrance of coriander while the flavour is gently aromatic. These fruits produce roughly 5% ash and also enclose tannin and malic acid. Chemical analysis of coriander has revealed that it also has high amounts of minerals as well as vitamins A, B and C. As coriander hardly contains any calorie, dieters prefer this herb (Rajeshwari and Andallu 2011).

Coriander is a commonly consumed green leafy vegetable (GLV) and occupies a prominent position in Indian culinary preparation. GLVs are rich sources of ascorbic acid, carotenoids, folic acid, riboflavin, vitamins and minerals such as phosphorous, iron and calcium (Prakash and Pal 1991). They are also known for their distinctive flavour, colour and therapeutic properties. Fruits and vegetables contribute around 95% of the total β-carotene available from all sources in India of which GLV provide almost 52% (Singh et al. 2001). Increasing the daily dietary intake of β-carotene, dietary fibre, as well as micronutrients through GLV, may help in attaining nutritional security, eradicate micronutrient deficiency and prevent degenerative diseases. GLV are comparatively economical and easily accessible in the market and are rich in iron and β-carotene which are vital for human health. The dietary approach to fighting micronutrient malnutrition is necessary for its role in increasing the consumption and availability of micronutrient-rich foods (Singh et al. 2001). Baskarachary (1996) reported that age-related macular degeneration (ARMD) and Vitamin A deficiency (VAD) is commonly caused due to the inadequacy of macular and provitamin A pigments in the Indian diet. The utilisation of carotenoid (zeaxanthin and lutein) rich foods lowers ARMD, cancer rate, cataract formation and cardiovascular disease (Landrum and Bone 2001; Wisniewska and Subczynski 2006). The green foliage is utilised for chutney, salads dressing and soup preparations (Ilyas 1980). The fresh green leaf is mostly used in cooking, and local varieties are known for the high production of foliage in the Transcaucasian area (Diederichsen 1996). The entire plant, i.e., leaves, fruits and stems of coriander have pleasant aromatic odour hence used as a flavouring substance. The complete plant is used in chutney preparation and leaves are utilised for flavouring curries, sauces and soups. Oleoresins and essential oil obtained from coriander are used in beverages, candy, canned soup, curry mix, condiments, chewing gums, baked goods, ice creams, gin, tobacco goods, sauces sweets, seasoning for various meat products, etc. (Rajeshwari and Andallu 2011). Coriander is one of the most utilised spice and forms an essential ingredient of various types of culinary preparations, chutneys, tomato sauces and salsa. It builds a staple spice in multiple cuisines in Central America, South America, Mexico, East India and Asia. The average utilisation of essential oil varies from 0.1 to 100 ppm. The culinary applcations of coriander include-

- Mostly used in chutneys, curries, pickles and sauces preparations
- Coriander has been utilised for the production of hygienic shelf-stable green dried foliage
- Pani puri spice mix from dried coriander foliage and coriander paste from fresh foliage
- Sprinkle over fruits and salads
- Utilised for flavouring curries, sauces and soups
- Used as a flavouring in bread, scones and waffles
- As an ingredient in gingerbread, biscuits and cake preparations
- As a flavouring in mayonnaise or creamed cheese
- Use to garnish fish, meat or vegetable dishes
- Used extensively in almost all homemade dishes, vegetable burgers, non-vegetable burgers and meatballs

- Added to all types of gravies and stocks
- Used in marinating meat and fish
- Coriander suits very well with mushrooms
- Used for spicing up spinach and stir-fries

MEDICINAL AND PHARMACOLOGICAL PROPERTIES

The pharmaceutical properties of coriander are well documented in Greek and Sanskrit writings. Hippocrates used this herb for its health benefits. Cilantro has an excellent hypoglycemic effect hence it is used as an anti-diabetic herb in some parts of Europe. Indian traditional medicine utilises coriander in urinary, respiratory and digestive disorders as it demonstrates diuretic, stimulant, diaphoretic and carminative effects. Coriander is used for its anti-inflammatory activity in India (Rajeshwari and Andallu 2011). It can also be utilised in treating diabetes, headaches, muscle pain, stiffness and lowering cholesterol level. Iranian traditional medicine employs coriander to get relief from insomnia, dyspeptic problems, loss of appetite, convulsion and anxiety. The different phytonutrient present in coriander possesses various health benefits including antibacterial, antimutagenic, antispasmodic, carminative, cytotoxic, fungicidal, hypoglycemic, hypolipidemic, insecticidal, lipolytic, stimulant, and stomachic activity (Benjumea et al. 2005; Maghrani et al. 2005; Mir 1992; Zargari 1991; Duke 2002; Chithra and Leelamma 1997; Dhanapakiam et al. 2008).

The essential oil extracted from coriander seed possess various therapeutic properties such as aperitif, carminative, aids in the nervous system and helps in digestion in intestine and stomach. Earlier, coriander leaves were used to treat erysipelas caused due to *Streptococcus* infection. Asians use foliage to cure enlargements, headaches and piles while the seed is used to treat colic, conjunctivitis and piles. The essential oil applied to cure colic, neuralgia and rheumatism while fruits in the form of a paste are utilised to cure various types of ulcers including mouth ulcer (Diederichsen 1996). Recent studies have confirmed coriander as a soother for the stomach in colicky infants and adults and help in healing diarrhoea, flatulence as well as indigestion. Antioxidants present in cilantro helps in delaying the decay of animal fats. This herb also contains specific compounds that prevent wound infections as well as destroys the bacteria and fungi causing meat spoilage. This herb alleviates pain and induces sleep in the infants.

Diluted tea prepared from coriander helps in reducing colic infection in children below two years. Seeds are used as a drug against worms, indigestion, arthritic pain, anti-inflammatory and rheumatism (Bisset 1994). Dry coriander seeds are highly effective in treating diarrhoea and flatulence. Boiled coriander seeds are beneficial for women suffering from hormonal mood swings and heavy menstrual flow. The presence of high antioxidants protects the body from the damages caused by free radicals. Coriander seed has bacteriocidal, fungicidal, anthelmintic property, alleviate abdominal pain and reduces digestive spasms. Seeds are the rich source of amino acids, proteins and fatty acids. Dried seeds are used to cure conjunctivitis. Coriander seed along with milk and honey reduces fever.

Coriander leaves contain a high amount of carotenoids. Fresh coriander leaf juices contain two-fold higher beta-carotene as compared to broccoli juice. Leaves are the powerful aphrodisiac and used to cure arthritis. Essential oil stimulates creativity, imagination and optimism. Citronellol present in coriander shows high antiseptic activity. The antifungal and antioxidant properties present in cilantro is utilised to treat eczema and various skin diseases including skin dryness. Ascorbic acid and iron also help to reduce pain and improve the immune system. Studies have also demonstrated its hypoglycemic action and effects on carbohydrate metabolism (Craig 1999; Chithra and Leelamma 2000).

The antimicrobial substances present in coriander helps in the treatment and prevention of smallpox as well as fights against *Salmonella* thus preventing food poisoning. The analgesic and heating effect present in coriander helps to cure rheumatism and pains in bone. Flavonoids in leaves are used to cure haemorrhoids and varices. Coriander also reduces hypertension and heals mouth sores and ulcers. It decreases the accumulation of heavy metals in the body. Thus, in turn, prevents memory loss and Alzheimer's disease. Studies have indicated that coriander contains insulin-like activity, muscle relaxant and sedative effects. Coriander also cools the body during the summer season. It also reduces irritation and eye burning in persons suffering from conjunctivitis. Coriander is utilised to get relief from insomnia and anxiety in Iran. Studies in mice confirmed its anxiolytic activity (Emamghoreishi et al. 2005).

Coriander leaves are an excellent source of iron, polyphenols, flavonoids and phytonutrients. It helps in digestion, lowers blood sugar level, prevents nausea and protects against urinary tract infections. Coriander juice is used in treating colitis, dysentery, indigestion and hepatitis. It acts as a potent remedy against pimples and blackheads when mixed with turmeric powder. It also has sedative and anxiolytic effects hence used to cure panic attacks, anxiety and depression. Linalool present in the seed helps to detoxify the liver and increase appetite. Table 4 shows various bioactive compounds identified in coriander leaves and their pharmacological properties (Ganesan et al. 2013).

Table 4. Pharmacological activity of various bioactive compounds from coriander leaves

Pharmacological activities	Bioactive compounds	Reference
Antioxidant	Apigenin; Ascorbic acid; Beta-carotene; Caffeic acid; Camphene	Ganesan et al. 2013; Maheswari and Cholarani 2013
Antimicrobial	Alpha-phellandrene; Alpha-pinene; Alpha-terpinene; Apigenin	Ganesan et al. 2013; Darughe et al. 2012; Kubo et al. 2004; Silva 2011; Al-Mofleh 2006
Antidiabetic	Chlorogenic acid; Ascorbic acid	Ganesan et al. 2013; Eidi et al. 2009; Gray and Flatt 1999
Anticholesterol	Crude extracts from coriander seeds and leaves	Ganesan et al. 2013; Chithra and Leelamma 1997
Anticancer	Alpha-pinene; Apigenin	Ganesan et al. 2013; Chithra and Leelamma 2000
Hepatoprotective	Phenolic compounds, quercetin and iso-quercetin	Ganesan et al. 2013; Pandey et al. 2011

Polyphenols are secondary metabolites possessing antihepatotoxic, anti-inflammatory, antimicrobial, antioxidant and antitumor etc. activities (Hertog 1995; Rice et al. 1996; Middleton et al. 2000). These secondary metabolites are considered as natural antioxidants and are present in various vegetables thus these vegetable acts as a functional food (McDonald et al. 2001).

Melo et al. (2005) reported that phenolic acid such as caffeic acid, protocatechinic acid and glycitin present in coriander aqueous extract exhibit high antioxidant activity. Juhaimi and Ghafoor (2011) have reported that different parts from parsley, mint and coriander vary in their phenolic content leading to different antioxidant activity.

Anti-Anthelmintic Activity

In-vitro anthelmintic activity of hydro-alcoholic and crude aqueous extracts from coriander seed has been carried out on adult and egg of *Haemonchus contortus*. *In-vivo* anthelmintic activity also has been studied in sheep infected with *Haemonchus contortus*. The results indicate that both hydro-alcoholic and crude aqueous extracts at less than 0.5 mg/mL concentration entirely repressed hatching of eggs (Eguale et al. 2007). This is attributed to the presence of alkaloids, flavonoids, polyphenols, phytosteroides and withanoids in crude hydro-alcoholic and aqueous extracts from coriander seed. They reported *in vitro* anthelmintic property of hydro-alcoholic and crude aqueous extracts of coriander seeds. Aqueous extracts induced 45% mortality while hydro-alcoholic extract showed 85% mortality rate at 0.5 mg/mL concentartion. The hatching of eggs was completely inhibited at a concentration of less than 0.5 mg/mL. The faecal egg count reduction test confirmed a significant reduction in egg per gram of faeces when infected sheep were administered with coriander extarct at a dose of 0.90 g/kg. A significant reduction in male worm count and total worm count was also reported for a dose of 0.90 g/kg of criander. Thus, hydro-alcoholic and crude aqueous extracts from seeds of *Coriandrum sativum* showed in vitro and in vivo anthelmintic activities against *Haemonchus contortus*.

Anti-Anxiety Effect

Anxiety is generally irritating and unmanageable internal disorder usually associated with nervous behaviour and abdomen ache etc. Coriander has been utilized as a folk medicine for treatment of sleeping disorder in Iran. Hydro-alcoholic extracts from coriander have been reported to demonstrate anti-anxiety activity in mice (Mahendra and Bisht 2011). This study confirmed that hydro-alcoholic extracts from coriander at 100 and 200 mg/kg concentration demonstrated anti-anxiety effects in mice similar to diazepam (a synthetic drug clinically used to treat anxiety). Hydro-alcoholic (70%) extract from coriander seed was explored for anti-anxiety activity using different animal models of anxiety (social interaction test, light and dark test, open field test, and elevated plus maze) in Swiss albino male mice (Mahendra and Bisht 2011). Hydroalcoholic extract from coriander seed at doses of 100 mg/kg and 200 mg/kg exhibited a similar anti-anxiety effect similar to standard diazepam. Administration of Hydro-alcoholic (70%) extract from coriander seed at doses of 100 mg/kg and 200 mg/kg significantly increased the percentage of open arm entries, the time spent in the open arms, the time spent in light compartment, and the time spent in social interaction in the male Swiss albino mice. The hydro-alcoholic extract also caused a significant reduction in the number of squares crossed in the perimeter and frequency of rearing in the mice. Thus results of this study confirmed the anti-anxiety potential of hydro-alcoholic extract from coriander (Mahendra and Bisht 2011).

Coriander also showed an anxiolytic effect in male albino mice when administered with 10, 25, 50, 100 mg/kg of aqueous extract. Neuromuscular coordination and spontaneous activity in mice reduced considerably when fed with aqueous coriander extract at 50, 100 and 500 mg/kg. The study confirmed sedative and muscle relaxant potential of coriander extract (Emamghoreishi et al. 2005). Administration of aqueous extract of coriander seed at a dose of 100 mg/kg significantly increased both the percentage of the open arm and the time spent in the open arms in the male albino mice using elevated plus-maze model. The aqueous extract of coriander seed caused a significant reduction in the spontaneous activity of male albino mice, reduction in the number of seconds spent on the rotarod and profound influence on motor coordination in a dose-dependent manner. Thus results of this study confirmed sedative and muscle relaxant potential of aqueous extract from coriander (Emamghoreishi et al. 2005).

Anti-Carcinogenic Effect

Coriander is a rich source of phthalides and found to exhibit substantial anti-cancerous property in different cell lines. Phthalides are also found to show the potential anticancer effect. Crude coriander extract is also reported to exhibit an efficient antitumor property against colon cancer. Coriander seed was reported to show protective action against 1,2-dimethylhydrazine (DMH) induced colon cancer in male albino rats. This study also confirmed the concentrations of phospholipids increase significantly while cholesterol to phospholipid ratio and the levels of cholesterol decreased significantly in the control group in comparison with the coriander seed administered group. Coriander seed-fed male albino rats also showed a sharp increase in bile acids, faecal dry weight and faecal neutral sterols as compared to DMH administered rats (Chithra and Leelamma 2000).

Antidiabetic and Anti-Hyperglycaemic Activity

Coriander leaves and seeds demonstrate a significant antidiabetic activity by facilitating the increased release of insulin from pancreatic cells. Thus, including coriander in drinking water or diet is an efficient way to reduce hyperglycemia. Administration of ethanolic extract from coriander seed at doses of 200 and 250 mg/kg to male Wistar rats showed a significant increase in the release of insulin from the beta cells and reduction in serum glucose level. Thus ethanolic extract from coriander seed exhibited a marked hypoglycaemic effect in streptozotocin-induced hyperglycaemic male Wistar rats (Eidi et al. 2009).

Coriander is utilized to cure diabetes in traditional medicine. Diabetic induced rats showed a significant rise in insulin level and a decline in blood glucose when fed with coriander seed extract. Diabetic induced rats inhibited peroxidative damage, reactivated antioxidant enzymes and antioxidant levels when administered with coriander seed extract. Coriander aqueous extract also exhibits insulin-like and insulin-releasing activity (Gray and Flatt 1999). Administration of coriander through drinking water (2.5 g/l) and diet (62.5 g/kg) to male streptozotocin-induced diabetic mice showed a significant decrease in hyperglycaemia, polydipsia and weight loss. Aqueous extract of coriander seed (1 mg/mL) increased glucose oxidation by 1.4-fold, glucose uptake by 1.6-fold and glycogenesis by1.7-fold in an isolated murine abdominal muscle. Aqueous extract of coriander seed also exhibited dose-dependent insulin secretion from a clonal β-cell lines. The action of aqueous coriander seed extract (1 mg/mL) was not enhanced by 1mM-3-isobutyl-1-methylxanthine, which increases cyclic AMP in insulin-secreting cells. In contrast, aqueous coriander seed extract further improved the insulin release by depolarised clonal β-cell lines. The results of this study display the presence of antihyperglycaemic, insulin-like and insulin-releasing activity in coriander. Coriander seed extract is utilized for diabetic patients as traditional medicine. This extract showed a marked decline in blood glucose levels in diabetic rats (Gray and Flatt 1999).

Deepa and Anuradha (2011) observed peroxidation damage inhibition in diabetic rats on the addition of coriander seed extract and also resulted in reactivation of antioxidant level and antioxidant enzymes in diabetic-induced rats.

Anti-Inflammatory Activity

Anti-inflammatory or anti-inflammation refers to the ability of a substance or treatment that reduces swelling or inflammation. Anti-inflammatory drugs are analgesics, which eradicates pain by reducing inflammation in contrast to opioids, which affect the central nervous system. Coriander is utilized to treat inflammation in traditional medicine. The ethanolic extracts of stem and leaf from coriander have been reported to exhibit a strong anti-inflammatory activity (Wu et al. 2010). The coriander extracts inhibit pro-inflammatory mediator expression by suppressing mitogen-activated protein kinases and NF-kappa-β activation. *In-vivo* study of lipolotion containing 0.5% coriander oil significantly reduced erythema in the ultraviolet (UV) erythema test thus suggesting its anti-inflammatory efficacy. Ethanolic (95%) extracts from coriander leaf and stem significantly decreased the production of prostaglandin E2 and lipopolysaccharide (LPS)-induced nitric oxide (Wu et al. 2010). Ethanolic extracts also reduced LPS-induced expression of phosphorylated mitogen-activated protein kinases (MAPKs), cyclooxygenase-2, nitric oxide synthase and pro-interleukin-1β expression. Ethanolic extracts from coriander leaf and stem dramatically inhibited the expression of nuclear p65 protein and LPS-induced IκB-α phosphorylation as well as reporter gene activity and NF-κB nuclear protein–DNA binding affinity. Thus it is evident from the study that ethanolic extracts from coriander leaf and stem exhibit strong anti-inflammatory activity as it inhibits MAPK signal transduction pathway and pro-inflammatory mediator expression by suppressing NF-κB activation in LPS-induced macrophages.

Antimicrobial Activity

A new molecule (Heneicos-1-ene) identified in coriander foliage also displayed significant antimicrobial activity against *Salmonella typhi* and *E. coli*. The antimicrobial activity of Heneicos-1-ene ranged between 25 to 50 μg/mL while it exhibited minimal bacterial concentration (MBC) between 50-100 μg/mL range (Priyadarshi et al. 2018).

Alkanals and aliphatic (2E)-alkenals isolated from fresh coriander leaves exhibited bactericidal activity against *Salmonella choleraesuis* spp. *choleraesuis* ATCC 35640. Fresh coriander leaves are reported to contain

(2E)-dodecenal and (2E)-undecenal which exhibit bactericidal activity against *Salmonella choleraesuis*. (2E)-Dodecenal possesses very effective bactericidal activity at a minimum bactericidal concentration (MBC) of 6.25 µg/mL while (2E)-undecenal showed bactericidal activity at MBC of 12.5 µg/mL. This α,β unsaturated aldehydes acts as nonionic surfactants which confer them with bactericidal activity at any growth stage (Kubo et al. 2004).

Kusuma et al. (2011) reported that coriander leaves exhibit antimicrobial activities against various bacteria and fungi. Coriander essential oil is reported to display high antimicrobial activity against various microorganisms. Dodecenal isolated from coriander exhibited two-fold higher inhibitory activity against *Salmonella* as compared to gentamicin (Kubo et al. 2004).

The main biotic factors for fungal deterioration of high moisture content food stuffs are pH and moisture content. Commonly reported moulds in bakery items and cakes are *Aspergillus niger, Monilia sitophila, Penicillium expansum, Penicillium stoloniferum, Rhizopus stolonifer* and species of *Geotrichum* and *Mucor*. The mycotoxins producing moulds are *Penicillium expansum, Mucor* and *Penicillium stoloniferum*. Darughe et al. (2012) demonstrated the antifungal property of coriander essential oil (CEO) in cakes. This study confirmed that there is no significant difference in the percentage of moulds after 30 days of storage in the cakes containing 0.01% butylated hydroxyanisole (BHA) and 0.05% CEO. It was also observed that 0.01% BHA and 0.05% CEO did not control mould growth, but CEO up to 0.15% showed better control on mould growth as compared to BHA and control samples. Coriander is utilized as a preventive and curative herb in spite of these chief functional properties. It also significantly reduces serum progesterone thus displaying antifertility activity. It is utilized in the preparation of domestic medicine to cure bed cold, nausea, seasonal fever, stomach disorders and vomiting. It is utilized as a drug against rheumatism, pain in the joints, worms and acidity. Coriander is often referred to as store house for bioactive compounds because it is rich in various phytonutrients which offer it different health benefits (Rajeshwari and Andallu 2011).

Coriander forms a protective layer over stomachal mucosal membranes thus preventing stomach against any ill effect. The aqueous suspension of coriander seed when administered at oral doses of 250 and 500 mg/kg body weight to male Wistar albino rats was found to form a protective layer on the surface of gastric mucosa thus protecting the cells from gastric mucosal injury. Different antioxidant compounds (catechin, coumarin, flavonoid, linalool, polyphenolic compounds and terpenoid) in the aqueous suspension of coriander seed might be responsible for the protective action against ethanol-induced gastric damage (Al-Mofleh 2006). The aqueous suspension was also reported to form dose-dependent protection against the (a) ethanol-induced histopathological lesions (b) ulcerogenic effects of ethanol, indomethacin, NaCl and NaOH; (c) pylorus ligated accumulation of gastric acid secretions and an ethanol-related decrease of Nonprotein Sulfhydryl groups (NP-SH).

Coriander oil shows a broad spectrum of antimicrobial activity against pathogenic fungus, gram-negative and gram-positive bacterium. This oil also exhibits a broad range of antiseptic activity. Coriander oil was reported to possess effective antimicrobial activity against *Acinetobacter baumannii*, *Bacillus cereus*, *Enterococcus faecalis*, *Streptococcus pneumoniae*, *Pseudomonas aeruginosa*, *Salmonella typhimurium*, *Staphylococcus aureus* and meticillin-resistant *Staphylococcus aureus* isolates. The treatment of bacterial cells with coriander oil causes damage to the membrane which ultimately leads to cell death (Silva 2011).

Antioxidant Activity

Oxidation is an essential step to fuel biological processes for the generation of energy. Oxygen is reduced to water during metabolism. The stepwise transfer of electrons during metabolism generates reactive oxygen species (ROS) such as hydrogen peroxide, hydroxyl and superoxide radicals. Other radicals formed during the metabolic activity are peroxyl, alkoxyl and alkyl radicals (Simic et al. 1989). The uncontrolled production of such ROS causes various diseases such as ageing, Alzheimer's disease, arteriosclerosis, cancer, coronary heart disease, diabetes mellitus, immune deficiencies, inflammation, arthritis, ischemia, neurologic disorders, Parkinson's disease and stroke (Sies 1991). Various compounds like tocopherols, glutathione and ascorbic acid, or enzymes such as catalase and superoxide dismutase protect the body from the free radicals damages. In an organism, there is equilibrium between the ROS generated and their removal. But during ageing, the production of ROS becomes higher than antioxidant protection leading to deterioration of physiological functions which finally causes accelerated ageing and diseases. Hence, phytoconstituents, medicinal plants and natural foods having high antioxidant power are gaining importance (Galvez et al. 2005). Coriander foliage is mainly consumed in India, Thailand, Singapore and Malaysia. Coriander foliage is a rich source of carotenoids, essential oil (linalool), polyphenols and other phytochemicals which contribute to its antioxidant activity (Maheswari and Cholarani 2013). Wangensteen et al. (2004) recorded coriander leaves to be a more potent antioxidant than the seeds. Oxidation damage is also controlled by radical scavenging activity mainly due to the presence of carotenoids which are also a potent antioxidant as reported by Peethambaran et al. (2012). Coriander essential oil (CEO) also acts as a potent natural antioxidant at a level of 0.05, 0.10 and 0.15. CEO is found to inhibit primary and secondary oxidation. Darughe et al. (2012) concluded that 0.02% of CEO was found to show similar activity to that exhibited by addition of equal amount of synthetic antioxidant (butylated hydroxyanisole).

It is also reported that the ethanolic extract of coriander foliage exhibits commendable radical scavenging activity (Priyadarshi et al. 2016). Further a new molecule exhibiting excellent radical scavenging activity was isolated and identified from coriander foliage. The new molecule (Heneicos-1-ene) was isolated via chromatographic technique, and its structure was established by employing multinuclei and multidimensional NMR and HRMS techniques (Priyadarshi et al. 2018). Heneicos-1-ene displayed a radical scavenging activity of 89.6±0.62% at 200 ppm levels.

Diuretic Activity

The diuretic activity of coriander extract was studied in the Wister rats of either sex (200-250 g). The positive control group was fed with furosemide (10 mg/kg), and the negative control group received saline while rest groups of animals received the different doses of coriander extracts dissolved in saline (50 mL/kg). The significant increase in diuresis confirmed the diuretic effect of coriander in rats. The diuresis results were found to be similar to the standard diuretic drug furosemide (Jabeen et al. 2009).

The diuretic activity of 70% aqueous-methanolic crude extract from coriander seed (Cs.Cr) was studied in the Wister rats of either sex (200-250 g). The positive control group were fed with furosemide (10 mg/kg), and the negative control group received saline while rest groups of animals received the different doses of coriander extracts dissolved in saline (50 mL/kg). Administration of Cs.Cr at a dose of 30 mg/kg resulted in a significant increase in urine output (diuresis) in rats. The significant increase in diuresis confirmed the diuretic effect of Cs.Cr in rats. The diuresis results were found to be similar to the standard diuretic drug furosemide (Jabeen et al. 2009).

Hepatoprotective Activity

Hepatotoxicity is the most commonly found disease in persons consuming alcohol for long-term. Patients mostly prefer natural medications consumed in daily foods to prevent hepatic failure. Coriander leaves are the rich source of bioactive compounds such as phenolic compounds, flavonoids and alkaloids. These compounds are responsible for the high hepatoprotective activity of coriander leaves. Coriander extract is found to exhibit higher hepatic protective with increasing enzymes and liver function (Pandey et al. 2011).

Treatment with CCl_4 causes in severe damage to the liver with the increased level of markers such as ALP, bilirubin, SGOT and SGPT in serum. Ethanolic (60%) extract from coriander leaves were reported to contain quercetin and iso-quercetin. Administration of ethanolic extract from coriander leaves before CCl_4 treatment resulted in a reduction in the liver weight as well as the levels of ALP, bilirubin, SGOT and SGPT in serum. Treatment with CCl_4 causes oxidative damage to the tissues and ethanolic extract from coriander leaves containing quercetin and iso-quercetin inhibits the generation of free radical thus exhibits hepatoprotective activity (Pandey et al. 2011).

Hypocholesterolemic Activity

Hypercholesterolemia is the major factor which leads to serious illness such as cancer, heart attack and hypertension. Natural foods are the most preferred choice for treating hypercholesterolemia. The crude extracts from coriander seeds and leaves are reported to have potent lipid-lowering property in rats. Coriander seeds, in contrast to coriander leaves, contain a higher amount of essential oils which is responsible for lipid-lowering effect with decreased triglycerides and cholesterol (Chithra and Leelamma 1997).

The levels of triglycerides, LDL+VLDL cholesterol and total cholesterol were reported to decrease significantly when female albino rats with high cholesterol level were administered with a diet containing coriander seeds (10%). Coriander seeds fed rats also showed a significant increase in HDL cholesterol, β-hydroxy, plasma lecithin-cholesterol acyltransferase and β-methyl glutaryl CoA reductase activity. Thus coriander seed was found to exhibit a significant hypolipidemic action which resulted in the increased degradation of cholesterol to neutral sterols and faecal bile acids, enhanced hepatic bile acid synthesis and increased activity of plasma LCAT in high cholesterol-fed female albino rats (Chithra and Leelamma 1997).

Hypolipidemic Effect

Hyperlipidaemia enhances the risk for the production of lipid oxidization product. These oxidized lipid products accumulate in the subendothelial areas of bone and vasculature. Atherogenic high-fat diets increase serum levels of oxidised lipids, which are best-known to attenuate osteogenesis in culture and to promote bone loss. Coriander showed enhanced lipid breakdown and decreased its uptake when fed to triton induced hyperlipidemic rats at a dose of 1g/kg. Thus, this study suggests that coriander is used as a curative and preventive herb against hyperlipidemia (Lal et al. 2004).

Metal Detoxification Activity

Coriander has the potential to remove toxic metals from the body hence it can be used as a natural cleaner. Coriander contains various compounds which bind with harmful metals and carry them away from the cells (Momin 2012).

Karunasagar et al. (2005) observed that coriander plant is very efficient to remove methylmercury (CH_3Hg^+) and inorganic (Hg^{2+}) from liquid solutions due to great binding efficacy of the carboxylic group to mercury. These results demonstrated that sorbent would be accustomed take away methylmercury (CH_3Hg^+) and inorganic (Hg^{2+}) from contaminated water. Column packed with silica-immobilized coriander also demonstrated that the sorbent is able to remove Hg^{2+} and CH_3Hg^+ from water. Carboxylic acid group plays the major role in binding the mercury.

CONCLUSION

The information provided in this chapter suggests that the coriander leaf and seeds provide important micro and macro nutrients and also they are good source of natural bioactive compounds having varied health benefits.

Coriander seed is a rich source of various minerals. Coriander is free from cholesterol, Vitamin B_{12} and Vitamin D. It also contains high amounts of linolenic and linoleic acids. Coriander foliage has a unique aromatic flavour and is a rich source of beta-carotene, folic acid, vitamin A, vitamin C and minerals. Coriander foliage has a wide range of applications in both food and pharmacological areas. Foliage contains various bioactive compounds such as alkaloids, flavonoids and polyphenols.

The various phytonutrients present in coriander are endowed with a wide range of pharmacological benefits such as stomachic, stimulant, lipolytic, insecticidal, hypolipidemic, hypoglycemic, hepatoprotective, fungicidal, diuretic diaphoretic cytotoxic, carminative, antispasmodic, antimutagenic, antibacterial, and aflatoxin protective potential. Oleoresins and essential oil obtained from coriander are used in various food industries. The distillation residues contain a high amount of proteins and fats thus making it suitable for animal feed. Thus, coriander is a natural source that can offer nutrition along with diverse health benefits.

REFERENCES

Al-Mofleh, I. A., Alhaider, A. A., Mossa, J. S., Al-Sohaibani, M. O., Rafatullah, S., and Qureshi, S. 2006. "Protection of gastric mucosal damage by *Coriandrum sativum* L. pretreatment in Wistar albino rats." *Environmental Toxicology and Pharmacology* 22:64-69. doi:10.1016/j.etap.2005.12.002.

Anitescu, G., Doneanu, C., and Radulescu, V. 1997. "Isolation of coriander oil: Comparison between steam distillation and supercritical CO_2 extraction." *Flavour and Fragrance Journal* 12:173-76. doi:10.1002/(SICI)1099-1026(199705)12:3<173::AID-FFJ630>3.0.CO;2-1.

Bandoni, A. L., Mizrahi, I., and Juarez, M. A. 1998. "Composition and quality of the essential oil of coriander (*Coriandrum sativum* L.) from Argentina." *Journal of Essential Oil Research* 10:581-584. doi:10.1080/10412905.1998.9700977.

Baskarachary. 1996. "*Studies on carotenoids in some plant foods as a source of vitamin A.*" PhD. Thesis submitted to National Institute of Nutrition, ICMR, Hyderabad, India.

Benjumea, D., Abdala, S., Luis, F. H., Paz, P. P., and Herrera, D. M. 2005. "Diuretic activity of *Artemisia thuscula*, an endemic canary species." *Journal of Ethnopharmacology* 100:205-209. doi: 10.1016/j.jep.2005.03.005.

Bisset, N. G., and Wichtl, M. 1994. "*Herbal Drugs and Phytopharmaceuticals.*" Medpharm GmbH Scientific Publishers, Stuttgart, CRC Press, Boca Raton: 91-95.

Carrubba, A., Torre, R., Di Prima, A., Saiano, F., and Alonzo, G. 2002. "Statistical analysis on the essential oil of Italian coriander (*Coriandrum sativum* L.) fruits of different ages and origins." *Journal of Essential Oil Research* 14:389–396. doi: 10.1080/10412905.2002.9699899.

Chaudhry, N. M. A., and Tariq, P. 2006. "Bactericidal activity of black peeper, bay leaf and coriander against oral isolates." *Pakistan Journal of Pharmaceutical Sciences* 19:214-218.

Chithra, V., and Leelamma, S. 1997. "Hypolipidemic effect of coriander seeds (*Coriandrum sativum*): mechanism of action." *Plant Foods for Human Nutrition* 51:167-172. doi:10.1023/A:1007975430328.

Chithra, V., and Leelamma, S. 2000. "*Coriandrum sativum*-effect on lipid metabolism in 1,2-dimethyl hydrazine induced colon cancer." *Journal of Ethnopharmacology* 71:457-463. doi: 10.1016/S0378-8741(00)00182-3.

Craig, W. J. 1999. "Health-promoting properties of common herbs." *The American Journal of Clinical Nutrition* 70:491S–499S. doi: 10.1093/ajcn/70.3.491s.

Darughe, F., Barzegar, M., and Sahari, M. A. 2012. "Antioxidant and antifungal activity of Coriander (*Coriandrum sativum* L.) essential oil in cake." *International Food Research Journal* 19:1253-1260.

Deepa, B., and Anuradha, C. V. 2011. "Antioxidant potential of *coriander sativum* L. seed extract." *Indian Journal of Experimental Biology* 49:30-38.

Dhanapakiam, P., Joseph, J. M., Ramaswamy, V. K., Moorthi, M., and Kumar, A. S. 2008. "The cholesterol lowering property of coriander seeds (*Coriandrum sativum*): Mechanism of action." *Journal of Environmental Biology* 29:53-56.

Diederichsen, A. 1996. "*Coriander (Coriandrum sativum L.). Promoting the conservation and use of underutilized and neglected crops 3.*" Institute of Plant Geneticsand Crop Plant Research, Gatersleben/ International Plant Genetic Resources Institute, Rome:1-82.

Duke, J. A. 2002. "*Handbook of Medicinal Herbs.*" Second edition, CRC Press LLC, Boca Raton, Florida, USA: 222-223.

Eguale, T., Getachew, T., Debella, A., and Desta, A. F. 2007. "*In vitro* and *in vivo* anthelmintic activity of crude extracts of *Coriandrum sativum* against *Haemonchus contortus.*" *Journal of Ethnopharmacology* 110:428-433. doi: 10.1016/j.jep.2006.10.003.

Eidi, M., Eidi, A., Saeidi, A., Molanaei, S., Sadeghipour, A., Bahar, M., and Bahar, K. 2009. "Effect of coriander seed (*Coriandrum sativum* L.) ethanol extract on insulin release from pancreatic beta cells in streptozotocin-induced diabetic rats." *Phytotherapy Research* 23:404-406. doi: 10.1002/ptr.2642.

Emamghoreishi, M., Khasaki, M., and Aazam, M. F. 2005. "*Coriandrum sativum*: evaluation of its anxiolytic effect in the elevated plus-maze." *Journal of ethanopharmacology* 96:365–370. doi: 10.1016/j.jep.2004.06.022.

Galvez, M., Martin, C. C., Houghton, P. J., and Ayuso, M. J. 2005. "Antioxidant activity of methanol extracts obtained from Plantago species." *Journal of Agricultural and Food Chemistry* 53:1927-1933. doi: 10.1021/jf048076s.

Ganesan, P., Phaiphan, A., Murugan, Y., and Baharin, B. S. 2013. "Comparative study of bioactive compounds in curry and coriander leaves: An update." *Journal of Chemical and Pharmaceutical Research* 5:590-594.

Gray, A. M., and Flatt, P. R. 1999. "Insulin-releasing and insulin-like activity of the traditional anti-diabetic plant *Coriandrum sativum* (coriander)." *British Journal of Nutrition* 81:203-209. doi:10.1017/S0007114599000392.

Greuter, W. 1994. "International Code of Botanical Nomenclature (Tokyo Code)." Adopted by the *Fifteenth International Botanical Congress*, Yokohama, August - September 1993. Koeltz Scientific Books, Konigstein.

Grieve, M. 1971. "*Coriander. A Modern Herbal: The Medicinal, Culinary, Cosmetic and Economic Properties, Cultivation and Folk-Lor*," volume 1. Dover Publications, New York: 221–222.

Hertog, M. G. L. 1995. "Flavonoid intake and long-term risk of coronary heart disease and cancer in the seven countries study." *Archives of Internal Medicine* 155:1184-1195. doi:10.1001/archinte.155.4.381.

Ilyas, M. 1980. "Spices in India 3." *Economic Botany* 34:236-259.

Jabeen, Q., Bashir, S., Lyoussi, B., and Gilani, H. 2009. "Coriander fruit exhibits gut modulatory, blood pressure lowering and diuretic activities." *Journal of Ethnopharmacology* 122:123-130. doi:10.1016/j.jep.2008.12.016.

Juhaimi, F., and Ghafoor, K. 2011. "Total phenols and antioxidant activities of leaf and stem extracts from coriander, mint and parsley grown in Saudi Arabia." *Pakistan Journal of Botany* 43:2235-2237.

Kansal, L., Sharma, V., Sharma, A., Lodi, S., and Sharma, S. H. 2011. "Protective role of *Coriandrum sativum* (coriander) extracts against lead nitrate induced oxidative stress and tissue damage in the liver and kidney in male mice." *International Journal of Applied Biology and Pharmaceutical Technology* 2:65-83.

Kansal, L., Sharma, A., and Lodi, S. 2012. "Potential health benefits of coriander (*Coriandrum sativum*): an overview." *International Journal of Pharmaceutical Research and Development* 4:10-20.

Karunasagar, D., Krishna, M. V., Rao, S. V., and Arunachalam, J. 2005. "Removal and preconcentration of inorganic and methyl mercury from aqueous media using a sorbent prepared from plant *Coriander sativum*." *Journal of Hazardous Material* 118:133-139. doi:10.1016/j.jhazmat.2004.10.021.

Kubo, I., Fujita, K., Kubo, A., Nihei, K., and Ogura, T. 2004. "Antibacterial activity of coriander volatile compounds against *Salmonella choleraesuis.*" *Journal of Agricultural and Food Chem*istry 52:3329-3332. doi:10.1021/jf0354186.

Kumar, A., Singh, R., and Chhillar, R. 2008. "Influence of omitting irrigation and nitrogen levels on growth, yield and water use efficiency of coriander (*Coriandrum sativum*)." *Journal Acta Agronomica Hungarica* 56:69-74. doi:10.1556/AAgr.56.2008.1.7.

Kusuma, I. W., Kuspradini, H., Arung, E. T., Aryani, F., Min, Y. H., and Kim, J. S. 2011. "Biological activity and phytochemical analysis of three indonesian medicinal plants, *Murraya koenigii, Syzygium polyanthum* and *Zingiber purpurea.*" *Journal of Acupuncture and Meridian Studies* 4:75-79. doi:10.1016/S2005-2901(11)60010-1.

Lal, A. A., Kumar, T., Murthy, P. B., and Pillai, K. S. 2004. "Hypolipidemic effect of *Coriandrum sativum* in triton-induced hyperlipidemic rats." *Indian Journal of Experimental Biology* 42:909-912.

Landrum, J. T., and Bone, R. A. 2001. "Lutein, zeaxanthin, and the macular pigment." *Archives of Biochemistry and Biophysics* 385:28–40. doi:10.1006/abbi.2000.2171.

Maghrani, M., Zeggwagh, N. A., Haloui, M., and Eddouks, M. 2005. "Acute diuretic effect of aqueous extract of *Retama raetam* in normal rats." *Journal of Ethnopharmacology* 99:31-35. doi:10.1016/j.jep.2005.01.045.

Mahendra, P., and Bisht, S. 2011. "Anti-anxiety activity of *Coriandrum sativum* assessed using different experimental anxiety models." *Indian Journal of Pharmacology* 43:574-577. doi:10.4103/0253-7613.84975.

Maheswari, N. U., and Cholarani, N. 2013. "Pharmacognostic effect of leaves extract of *Murraya koenigii* Linn." *Journal of Chemical and Pharmaceutical Research* 5:120-123.

Maroufi, K., Farahani, H. A., and Darvishi, H. H. 2010. "Importance of Coriander (*Coriandrum sativum* L.) between the medicinal and aromatic plants." *Advances in Environmental Biology* 4:433-436.

McDonald, S., Prenzler, P. D., Antolovich, M., and Robards, K. 2001. "Phenolic content and antioxidant activity of olive extracts." *Food Chemistry* 73:73-84. doi:10.1016/S0308-8146(00)00288-0.

Melo, E. D., Filho, J. M., and Guerra, N. B. 2005. "Characterization of antioxidant compounds in aqueous coriander extract (*Coriandrum sativum* L.)." *LWT-Food Science and Technology* 38:15-19. doi:10.1016/j.lwt.2004.03.011.

Middleton, E., Kandaswami, C., and Theoharides, T. C. 2000. "The effects of plant flavonoids on mammalian cells: implications for inflammations, heart disease, and cancer." *Pharmacological Reviews* 52:673-751.

Mir, H. H. 1992. "*Coriandrum sativum* In: Application of plants in prevention and treatment of illnesses." *Islamic Cultural Publication, Tehran* 1:247-252.

Momin, A. H., Acharya, S. S., and Gajjar, A. V. 2012. "*Coriandrum sativum*: Review of advances in psychopharmacology." *International Journal of Pharmaceutical Sciences and Research* 3:1233-1239. doi:10.13040/IJPSR.0975-8232.3(5).1233-39.

Msaada, K., Jemia, M. B., Salem, N., Bachrouch, O., Sriti, J., Tammar, S., Bettaieb, I., Jabri, I., Kefi, S., Limam, F., and Marzouk, B. 2017. "Antioxidant activity of methanolic extracts from three coriander (*Coriandrum sativum* L.) fruit varieties." *Arabian Journal of Chemistry* 10:S3176-S3183. doi:10.1016/j.arabjc.2013.12.011.

Nambiar, V. S., Daniel, M., and Guin, P. 2010. "Characterization of polyphenols from coriander leaves (*Coriandrum sativum*), red amaranthus (*A. paniculatus*) and green amaranthus (*A. frumentaceus*) using paper chromatography: and their health implications." *Journal of Herbal Medicine and Toxicology* 4:173-177.

Pandey, A., Bigoniya, P., Raj, V., and Patel, K. K. 2011. "Pharmacological screening of *Coriandrum sativum* Linn. for hepatoprotective activity." *Journal of Pharmacy & Bio Allied Sciences* 3:435-441.

Parthasarathy, V. A., Chempakam, B., and Zachariah, T. J. 2008. "*Coriander in chemistry of spices*." CAB International UK 190-206.

Pathak, N. L., Kasture, S. B., Bhatt, N. M., and Rathod, J. D. 2011. "Phytopharmacological Properties of *Coriander sativum* as a Potential

Medicinal Tree: An Overview." *Journal of Applied Pharmaceutical Science* 1:20-25.

Peethambaran, D., Bijesh, P., and Bhagyalakshmi, N. 2012. "Carotenoid content, its stability during drying and the antioxidant activity of commercial coriander (*Coriandrum sativum L.*) varieties." *Food Research International* 45:342-350.

Prakash, D., and Pal, M. 1991. "Nutritional and antinutritional comparison of vegetable and grain *Amaranthus* leaves." *Journal of the Science of Food and Agriculture* 57:573–585. doi:10.1002/jsfa.2740570410.

Prakash, V. 1990. "*Leafy Spices.*" CRC Press Inc., Boca Raton: 31-32.

Priyadarshi, S., and Borse, B. B. 2014. "Effect of the environment on content and composition of essential oil in coriander." *International Journal of Scientific & Engineering Research* 5:57-65.

Priyadarshi, S., Khanum, H., Ravi, R., Borse, B. B., and Naidu, M. M. 2016. "Flavour characterisation and free radical scavenging activity of coriander (*Coriandrum sativum L.*) foliage."*Journal of Food Science and Technology* 53:1670-1678. doi:10.1007/s13197-015-2071-1.

Priyadarshi, S., Harohally, N. V., Roopavathi, C., and Naidu, M. M. 2018. "Isolation, identification, structural elucidation and bioactivity of Heneicos-1-ene from *Coriandrum sativum* L. foliage." *Scientific Reports* 8:1-6. doi:10.1038/s41598-018-35836-z.

Purseglove, J. W., Brown, E. G., Green, C. L., and Robbins, S. R. J. 1981. "*Spices.*" volume 2. Longman, New York:736–788.

Rajeshwari, U., and Andallu, B. 2011. "Medicinal benefits of coriander (*Coriandrum sativum* L)." *Spatula DD* 1:51-58. doi:10.5455/spatula.20110106123153.

Ramezani, S., Rahamanian, M., Jahanbin, R., Mo-hajeri, F., Rezaei, M. R., and Solaimani, B. 2009. "Diurnal Changes Essential oil content of Coriander (*Corian-drum sativum* L.) Aerial parts from Iran." *Research Journal of Biological Sciences* 4 (3): 277-281.

Rao, A. S., Ahmed, M. F., and Ibrahim, M. 2012. "Hepatoprotective activity of Meliaazedarach leaf extract against simvastatin induced Hepatotoxicity in rats." *Journal of Applied Pharmaceutical Science* 2:144-148. doi:10.7324/JAPS.2012.2721.

Rice, E. C. A., Miller, N. J., and Paganga, G. 1996. "Structure antioxidant activity relationships of flavonoids and phenolic acids." *Free Radical Biology and Meicine* 20:933-956. doi:10.1016/0891-5849(95)02227-9.

Sahib, N. G., Anwar, F., Gilani, A. H., Hamid, A. A., Saari, A., and Alkharfy, K. M. 2012. "Coriander (*Coriandrum sativum L.*): A potential source of high-value components for functional foods and nutraceuticals, A Review." *Phytotherapy Research* 27:1439-1456.

Sharma, M. M., and Sharma, R. K. 2012. "*Coriander, Handbook of herbs and spices*." Woodhead Publishing Ltd, United Kingdom: 216-249.

Sies, H. 1991. "Oxidative stress: from basic research to clinical application." *The American Journal of Medicine* 91:S31-S38. doi:10.1016/0002-9343(91)90281-2.

Silva, F., Ferreira, S., Queiroz, J. A., and Domingues, F. C. 2011. "Coriander (*Coriandrum sativum* L.) essential oil: its antibacterial activity and mode of action evaluated by flow cytometry." *Journal of Medical Microbiology* 60:1479–1486. doi:10.1099/jmm.0.034157-0.

Simic, M. G., Bergtold, D. S., and Karam, L. R. 1989. "Generation of oxy radicals in biosystems." *Mutation Research/Fundamental and Molecular Mechanisms of Mutagenesis* 214:3-12. doi:10.1016/0027-5107(89)90192-9.

Singh, G., Kawatra, A., and Sehgal, S. 2001. "Nutritional composition of selected green leafy vegetables, herbs and carrots." *Plant Foods for Human Nutrition* 56:359-364.

Singh, V. P., and Ramanujam, S. 1973. "Expression of an-dromonecy in coriander, *Coriandrum sativum* L." *Eu-phytica* 22:181-188.

Uhl, S. R. 2000. "*Coriander. Handbook of Spices, Seasonings, and Flavorings.*" Technomic Publishing Co., Inc., Lancaster: 94–97.

USDA. 2013. "*National Nutrient Database for Standard Reference Release 26 Full Report (all nutrients) Nutrient data for Spices, coriander seed.*"

Verma, A., Pandeya, S. N., Yadav, S. K.., Singh, S., and Soni, P. A. 2011. "A review on *Coriandrum sativum* (Linn.): An Ayurvedic Medicinal Herb of Happiness." *Journal of Advances in Pharmacy and Healthcare Research* 1:28-48.

Wallis, T. E. 2005. "*Textbook of Pharmacognosy*." 5th edtion S. K. Jain for CBS publishers and distibuters; New Delhi (India):125-248.

Wangensteen, H., Samuelsen, A. B., and Malterud, K. E. 2004. "Antioxidant activity in extracts from coriander." *Food Chemistry* 88:293–297. doi:10.1016/j.foodchem.2004.01.047.

Wisniewska, A., and Subczynski, W. K. 2006. "Distribution of macular xanthophylls between domains in a model of photoreceptor outer segment membranes." *Free and Radical Biology Medicine* 41:1257–1265. doi:10.1016/j.freeradbiomed.2006.07.003.

Wu, T. T., Tsai, C. W., Yao, H. T., and Lii, C. K. 2010. "Suppressive effects of extracts from the aerial part of *Coriandrum sativum* L. on LPS-induced inflammatory responses in murine RAW 264.7 macrophages." *Journal of the Science of Food and Agriculture* 90:1846-1854. doi:10.1002/jsfa.4023.

Zargari, A. 1991. "*Coriandrum sativum* L. *Herbal Medicine* 1:586–590.

In: Coriander
Editor: Deepak Kumar Semwal

ISBN: 978-1-53616-483-1
© 2019 Nova Science Publishers, Inc.

Chapter 2

DRIED CORIANDER: PROCESSING AND PROPERTIES

Raquel P. F. Guiné[1,*], *Sofia G. Florença*[2]
and Maria João Barroca[3,4]
[1]CI&DETS and CERNAS Research Centers,
Polytechnic Institute of Viseu, Viseu, Portugal
[2]Faculty of Nutrition Sciences, University of Porto,
Oporto, Portugal
[3]Molecular Physical-Chemistry, R&D Unit,
University of Coimbra, Coimbra, Portugal
[4] Coimbra College of Agriculture,
Polytechnic Institute of Coimbra, Coimbra, Portugal

ABSTRACT

Coriander leaves and fruits are rich in bioactive compounds, such as phenolic compounds, and most especially phenolic acids and flavonoids, and therefore exhibits high antioxidant capacity both in the leaves and the fruits. While coriander leaves contain important amounts of minerals and

---

[*]Corresponding Author's Email: raquelguine@esav.ipv.pt

vitamins, the seeds are valued mostly by their oil, with a fatty acids profile characterized by important contents of oleic, linoleic and petroselinic acids. Because the use of the fresh coriander sometimes is limited, processing operations are sued to extend the shelf life and increase availability and convenience. Processing operations include for example freezing or drying. This last, in particular, has many advantages, like the facility to use and store, but is also comprises some possible adverse effects of the quality of the treated products, with possible interference in the bioactivity of the chemical components present in the fresh form. There are many different drying methodologies, isolated or in combinations, that impart different levels of influence on the quality of the product. Therefore, the aim of the present chapter is to discuss the drying operation applied to coriander and evaluate the dried product's characteristics and to what extent they are affected as compared to the fresh herb.

Keywords: bioactive compounds, drying method, composition, processing

INTRODUCTION

Coriander (*Coriandrum sativum* L.), is grown both for its leaves as well as for the fruits, which bare a characteristic, spicy aroma and bitter flavour (Szempliński et al., 2018). Coriander has been used since ancient times for cooking, as a flavouring agent and as medication. Although it is a native species of the Mediterranean region, presently is also broadly grown, Central Europe, North Africa, Russia and Asia (Beyzi et al., 2017; Sahib et al., 2013).

Though all parts of the plant are edible, the fresh leaves and the dried seeds (the fruits) are those most used for culinary applications (Szempliński et al., 2018).

The fresh leaves are appreciated because they provide an exotic flavour to culinary preparations when added right before serving. Besides, they are also used in salads and as a garnish. Even though the roots and stems can be cooked in stews and soups, they are usually removed before serving (Guiné and Gonçalves, 2016). Although their organoleptic properties are more intense in fresh, the coriander leaves can be frozen or processed into the dried form for better preservation.

Coriander fruits are used as raw material for the extraction of an essential oil rich in several terpene compounds, containing 60–80% of linalool, which has demonstrated important antibacterial properties. Regarding the fatty oils, the coriander fruits contain about 20% of triglyceride (fatty oil), from which the main fatty acid is petroselinic acid, which is an isomer of the oleic acid. The methyl esters in coriander fatty oil show high oxidative stability and present better low-temperature properties as well as lower iodine value as compared with methyl ester from soybean oil (Khodadadi et al., 2016).

Some of the recognized medicinal properties of coriander fruits and essential oil include improvement of intestinal function, stimulation of digestion, treatment of rheumatism and relieve anxiety and tension (Khodadadi et al., 2016; Mandal and Mandal, 2015; Sayed-Ahmad et al., 2017). Also, coriander has been reported to exhibit numerous beneficial effects for the human health, namely antioxidant, anti-diabetic, anti-mutagenic, anthelmintic, sedative-hypnotic, anticonvulsant, diuretic, antifungal, anticancer, anxiolytic, hepatoprotective, anti-protozoal, anti-ulcer, post-coital, anti-fertility, cholesterol-lowering, protective against lead toxicity and heavy metal detoxifier (Beyzi et al., 2017; Momin et al., 2012).

Coriander essential oil is a natural antioxidant also possessing antifungal properties. Its utilization includes an addition to foods and especially those containing lipids where it will prevent lipid oxidation (Khodadadi et al., 2016; Mandal and Mandal, 2015; Sayed-Ahmad et al., 2017).

The fruits are used as spices owing to their unique flavour and preservation properties on stored foods. Other diverse applications of the fruits include the production of spirits, cosmetics, textiles, printing materials, and animal feeds (Khodadadi et al., 2016; Mandal and Mandal, 2015; Sayed-Ahmad et al., 2017). Coriander seeds can be utilized as a seasoning agent in liqueurs, teas, meat products and pickles (Beyzi et al., 2017).

COMPOSITION OF RAW CORIANDER

The coriander plant constitutes a rich reservoir of micronutrients and nutritional elements that will be also transferred to the fruits or seeds (Bhat et al., 2014). The seeds are mostly composed of essential oils, triglycerides, sugars, proteins and vitamin C (Beyzi et al., 2017). Tables 1, 2 and 3 report the chemical composition of coriander seeds and leaves in terms of major components, dietary minerals and vitamins, respectively.

Table 1. Proximate composition of fresh coriander

Component (g/100 g)	Seeds[1]	Leaves[1]
Water	8.86	92.21
Protein	12.37	2.13
Total lipid (fat)	17.77	0.52
Saturated fatty acids	*0.990*	*0.014*
Monounsaturated fatty acids	*13.580*	*0.275*
Polyunsaturated fatty acids	*1.750*	*0.040*
Trans fatty acids	—	*0.00*
Cholesterol	*0.00*	*0*
Carbohydrate	54.99	3.67
Total dietary fibre	*41.9*	*2.8*
Total sugars	—	*0.87*
Energy (kJ)	71	6
Energy (kcal)	298	23

[1] According to USDA Food Composition Databases (USDA, 2018).

According to Table 1, while the leaves are mostly composed of water (92%), the main components of seeds are carbohydrates (55%), lipids (18%) and proteins (12%). The seeds present high value of dietary fibre (42%) as compared with the leaves in which the fibre content is reduced to about 3%. Regarding the fatty acids profile, the monounsaturated fatty acids (MUFAs) are dominant both in the leaves and the seeds, although the relative amounts are much different, being about 14% in the seeds and only 0.3% in the leaves. This is so because the seeds have essential oil and therefore the total fatty acids amount present is expected to be high. The saturated fatty acids

(SFAs) are a minority in relation to unsaturated fatty acids, endowing these fatty acids profiles of beneficial health properties. Many studies have unequivocally demonstrated the harmful effects of SFAs on human health, and they are considered unhealthy with respect to cardiovascular diseases due to their negative effects on cholesterol metabolism, increasing the risks associated with metabolic factors, hypertension, myocardial infarction and oxidative stress. For this reason, several health authorities recommend limited ingestion of SFAs in the diet (Lichtenstein et al., 2006; Nakamura et al., 2019; Praagman et al., 2019). Contrarily, unsaturated fatty acids (MUFAs and PUFAs - polyunsaturated fatty acids) have demonstrated beneficial effects for the human health, like for example for bone health, preventing osteoporosis, as anti-microbial agents and as prodrugs for cancer therapy (Das, 2018; Kasonga et al., 2019; Sun et al., 2017).

In Table 2 are presented the content in dietary minerals of the seeds and leaves of coriander. The seeds appear as particularly rich in minerals as compared with the leaves, this being much owing to the high amount of water present in the leaves, that turn the concentrations of all other components much lower. Hence, the seeds constitute a good dietary source of calcium, iron, magnesium, phosphorus and potassium (respectively 709, 16, 330, 409 and 1267 mg/100 g). The importance of these minerals for good general health status is well known, for example, calcium for bone health or magnesium for muscles and mental health (as anti-depressive) (Schtscherbyna et al., 2019; Sun et al., 2019). Heavy metals like plumb, chromium or cadmium were evaluated by Beyzi et al. (2017) and the values obtained were very low (< 1 mg/100 g in all cases).

The amounts of vitamins present in coriander seeds and leaves are generally modest, as shown in Table 3. However, it is noteworthy to highlight the presentence of important amounts of vitamins A and K in the leaves, respectively 337 and 310 µg/100 g, as well as vitamin C in both the leaves and seeds (27 and 21 mg/100 g, respectively).

Table 2. Mineral composition of fresh coriander

Component	Seeds[1]	Seeds[2]	Leaves[1]
Calcium, Ca (mg/100 g)	709	1989 – 2772	67
Iron, Fe (mg/100 g)	16.32	12 – 25	1.77
Magnesium, Mg (mg/100 g)	330	244 – 307	26
Phosphorus, P (mg/100 g)	409	377 – 657	48
Potassium, K (mg/100 g)	1267	1879 – 2611	521
Sulfur, S (mg/100 g)	—	164 – 191	—
Boron, B (mg/100 g)	—	0.8 – 2.1	—
Sodium, Na (mg/100 g)	35	7 – 9	46
Cadmium, Cd (mg/100 g)	—	0.02 – 0.04	—
Chromium, Cr (mg/100 g)	—	0.07 – 0.12	—
Copper, Cu (mg/100 g)	—	1.7 – 2.0	—
Nickel, Ni (mg/100 g)	—	0.23 – 0.34	—
Manganese, Mn (mg/100 g)	—	4.6 – 8.3	—
Plumb, Pb (mg 100 g)	—	0.12 – 0.27	—
Zinc, Zn (mg/100 g)	4.70	2.7 – 7.8	0.50

[1] According to USDA Food Composition Databases (USDA, 2018).
[2] Range, depending on the variety, according to Beyzi et al. (2017).

Table 3. Vitamin content in fresh coriander (USDA, 2018)

Component	Seeds	Leaves
Vitamin A, retinol RAE[1] (µg/100 g)	0	337
Vitamin C, total ascorbic acid (mg/100 g)	21.0	27.0
Vitamin B_1, thiamin (mg/100 g)	0.239	0.067
Vitamin B_2, riboflavin (mg/100 g)	0.290	0.162
Vitamin B_3, niacin (mg/100 g)	—	1.114
Vitamin B_6, pyridoxin (mg/100 g)	—	0.149
Vitamin B_9, total folate (µg/100 g)	0	62
Vitamin B_{12}, cobalamin (µg/100 g)	0.0	0.0
Vitamin D, D2 + D3 (µg/100 g)	0.0	0.0
Vitamin E, alpha-tocopherol (mg/100 g)	—	2.50
Vitamin K, phylloquinone (µg/100 g)	13.2	310

[1] RAE = retinol activity equivalents.

Uitterhaegen et al. (2016) evaluated the composition of coriander seed oil reporting about 98% of triglycerides (TAG), 1% diglycerides (DAG), 0.1% of monoglycerides (MAG) and 1% of free fatty acids (FFA), with percentages expressed as petroselinic acid, for being the most abundant FA present. The amount of FFA present is an indicator of the quality of the oil.

Table 4 shows the composition of coriander seed essential oil. In the work by Beyzi et al., (2017) the major fatty acid found in coriander seed oil was petroselinic acid, representing 80-82%, depending on the coriander cultivar. After appeared linoleic acid, representing 14-15%, and then palmitic acid, 3 to 4%. The high content in petroselinic acid was also reported by Uitterhaegen et al. (2016), although lower than the previous study (about 73%). The anti-ageing and anti-inflammatory activities of petroselinic acid allow new application opportunities for the coriander seed oil linked to the cosmetic and functional food industries.

Table 4. Fatty acid composition of coriander seed oil

Component	Seed oil[1] (%)	Seed oil[2] (%)
Caproic acid: C6:0	—	0.1
Palmitic acid. C16:0	3.11 – 3.72	2.9
Palmitoleic acid: C16:1	—	< 0.1
Margaric acid: C17:0	—	—
Stearic acid: C18:0	0.70 – 1.66	< 0.1
Petroselinic acid: C18:1 $n12$	79.78 – 81.96	72.6
Oleic acid: C18:1 $n9$	—	6.0
Vaccenic acid: C18:1 $n7$	—	1.2
Linoleic acid: C18:2 $n6c$	13.51 – 14.72	13.8
Alpha-linolenic acid: C18:3 $n3$	0.20 – 0.27	0.2
Arachidic acid: C20:0	0.11 – 0.70	0.1
Saturated fatty acids: SFAs	—	—
Monounsaturated fatty acids: MUFAs	—	—
Polyunsaturated fatty acids: PUFAs	—	—
Crude oil content: COC	4.79 – 6.25	—

[1] Range, depending on the variety, reported by Beyzi et al. (2017).
[2] Reported by Uitterhaegen et al. (2016).

Table 5 presents the mineral composition of coriander oil as reported by different authors (Beyzi et al., 2017; Uitterhaegen et al., 2016). The results obtained in the reported works are much different in some cases, like for example for calcium and sodium, with much higher values reported by Beyzi et al. (2017), or for phosphorous and potassium, with higher values reported by Uitterhaegen et al. (2016).

Table 5. Mineral composition of coriander seed oil

Component	Seed oil[1]	Seed oil[2]
Calcium, Ca (mg/100 g)	426 – 619	9.28
Iron, Fe (mg/100 g)	0.02 – 0.05	0.14
Magnesium, Mg (mg/100 g)	7.4 – 9.9	3.58
Phosphorus, P (mg/100 g)	1.2 – 1.5	23.07
Potassium, K (mg/100 g)	0.8 – 1.2	7.31
Sulfur, S (mg/100 g)	6.0 – 6.7	—
Boron, B (mg/100 g)	0.12 – 0.21	—
Sodium, Na (mg/100 g)	127 – 201	0.52
Cadmium, Cd (mg/100 g)	0.01 – 0.03	—
Chromium, Cr (mg/100 g)	0.00 – 0.02	—
Copper, Cu (mg/100 g)	0.10 – 0.19	—
Nickel, Ni (mg/100 g)	0.09 – 0.17	—
Manganese, Mn (mg/100 g)	0.01 – 0.05	—
Plumb, Pb (mg 100 g)	0.00 – 0.02	—
Zinc, Zn (mg/100 g)	0.62 – 1.19	—

[1] Range, depending on the variety and extraction solvent, according to Beyzi et al. (2017).
[2] Reported by Uitterhaegen et al. (2016).

According to the results in Table 6, the coriander seed oil is a rich source of phytosterols, with a total amount of about 668 mg/100 g, making it one of the richest vegetable oils in terms of phytosterols, more than olive oil, for example (Uitterhaegen et al., 2016). However, the value reported by Kozłowska et al. (2016) was about half (347 mg/100 g). Phytosterols have attracted extensive attention due to their health benefits, namely in terms of preventing some chronic diseases related to unhealthy lifestyles, like, for example, obesity, type 2 diabetes, hyperlipidemia or hypertension.

Furthermore, plant sterols have proven inhibitory effects over the absorption of dietary and endogenously generated cholesterol from intestinal cells, reducing in this way the levels of circulating cholesterol. Because high serum cholesterol is among the major risk factors for cardiovascular disease, the ability of phytosterols to lower cholesterol absorption might contribute to an effective reduction of the risk of heart diseases (Mohamed, 2014; Vu et al., 2012).

Table 6. Sterol composition of coriander seed oil

Component	Seed oil[1]	Seed oil[2]
Cholesterol (mg/100 g)	2	—
Campesterol (mg/100 g)	54	39
Stigmasterol (mg/100 g)	161	—
Sitostanol (mg/1200 g)		27
β-Sitosterol (mg/100 g)	231	100
Δ^5-Avenasterol (mg/100 g)	27	52
Δ^5-Stigmasterol (mg/1200 g)	—	110
Δ^7-Stigmastenol (mg/100 g)	122	—
Δ^7-Avenasterol (mg/100 g)	40	14
Gramisterol (mg/100 g)	7	—
Citrostadienol (mg/100 g)	10	—
Cycloartenol (mg/100 g)	8	7
Methylene cycloartanol (mg/100 g)	6	—
Total sterols (mg/100 g)	668	347

[1] Reported by Uitterhaegen et al. (2016).
[2] Reported by Kozłowska et al. (2016).

The composition of tocopherols and tocotrienols was a reporter by Uitterhaegen et al. (2016) and is listed in Table 7. The global content of tocols is approximately 50 mg/100 g, being in the range for other vegetable oils, and have an important role to protect the oil against oxidative reactions. From the tocols present, the most abundant were γ-tocotrienol and α-tocotrienol, respectively with approximate contents of 35 and 10 mg/100 g. These tocols have been reported as lipophilic antioxidants (Zhang et al., 2013).

The total phospholipid content evaluated in coriander seed oil by Uitterhaegen et al. (2016) was 310 mg/100 g, from which phosphatidic acid was the most abundant subclass, representing 33% of all phospholipids. Next classes in importance are phosphatidylcholine (25%) and phosphatidylinositol and phosphatidylethanolamine (each one representing 17%). The phospholipid content of coriander oil is low as compared with other oils (only 0.31%) and may constitute an advantage when further refining is sought, especially for degumming, since a higher content in phospholipids increases the need for further refining (Shahidi, 2005).

Table 7. Tocols composition of coriander seed oil

Component	Seed oil[1]
α-tocopherol (mg/100 g)	1.24
α-tocotrienol (mg/100 g)	9.80
β-tocopherol (mg/100 g)	—
γ-tocopherol (mg/100 g)	1.01
γ-tocotrienol (mg/100 g)	35.03
δ-tocopherol (mg/100 g)	—
δ-tocotrienol (mg/100 g)	2.57
Total tocols (mg/100 g)	49.65

[1] Reported by Uitterhaegen et al. (2016).

Table 8. Phospholipid composition of coriander seed oil

Component	Seed oil[1]
Phosphatidic acid (mg/100 g)	100.8
Phosphatidylcholine (g/100 g)	78.7
Phosphatidylinositol (g/100 g)	52.7
Phosphatidylethanolamine (g/100 g)	51.8
Phosphatidyl glycerol (g/100 g)	25.1
1-lysophosphatidylcholine (g/100 g)	1.6
Total phospholipids (mg/100 g)	310.7

[1] Reported by Uitterhaegen et al. (2016).

Table 9 is summarized the total phenolic compounds determined by different authors and in different parts of the coriander plant, namely leaves, seeds and seed oil. The values were, when possible, converted to similar or equal units for better understanding of the different results obtained. The leaves demonstrated the highest content in phenolic compounds (3.6-54.5 mg GAE/g) when compared with the seeds (0.9-18.9 mg GAE^2/g) or the oil, this last showing the lowest value of all (0.17-0.20 mg GAE/g).

Table 9. Phenolic compounds in coriander parts

Coriander part	Total phenolic compounds
Leaves [1] (mg GAE^2/g)	3.6 – 54.5
Seeds[1] (mg GAE^2/g)	0.9 – 18.9
Seed oil[3] (mg GAE^2/g)	1.4
Seed oil[4] (mg GAE^2/mL)	0.32 – 0.51
Seed oil[5] (mg GAE^2/g)	0.17 – 0.20

[1] Range, depending on the extraction solvent, reported by Wangensteen et al. (2004).
[2] GAE = Galic Acid Equivalent.
[3] Value reported by Wangensteen et al. (2004).
[4] Range, depending on cultivar, reported by Beyzi et al. (2017).
[5] Range, depending on the extraction solvent, reported by Kozłowska et al. (2016).

Table 10. Antioxidant activity in coriander parts

Property	Coriander part	value	Reference
ORAC[1] antioxidant capacity (mmol TE^2/100 g dm^3)	Aerial parts (leaves and stems)	2.85	El-Zaeddi et al. (2017)
FRAP[4] antioxidant capacity (mmol TE^2/100 g dm^3)	Aerial parts (leaves and stems)	4.60	El-Zaeddi et al. (2017)
ABTS[5] antioxidant capacity (mmol TE^2/100 g dm^3)	Aerial parts (leaves and stems)	1.55	El-Zaeddi et al. (2017)
DPPH[6] antiradical activity (% inhibition)	Seed oil	66.7-76.8	Beyzi et al. (2017)
DPPH[6] antiradical activity (μmol TEAC/g)	Seed oil	2.24-4.96	Kozłowska et al. (2016)

[1] ORAC = Oxygen Radical Absorbance Capacity.
[2] TE = Trolox equivalent.
[3] dm = dry mass.
[4] FRAP = Ferric Reducing Antioxidant Power Assay.
[5] ABTS = [2,2'-azinobis-(3-ethylbenzothiazoline-6-sulfonate)] Assay.
[6] DPPH = [1,1-Diphenyl-2-picryl-hydrazyl] Antioxidant Assay.

The results in Table 10 refer to the antioxidant activity of different parts of coriander, determined by different analytical methods, and therefore, not comparable, not even for the two values obtained by DPPH method in the seed oil, because they were expressed in different units by the authors. Hence, it is not possible to conclude about the relative antioxidant capacity of the coriander parts, although it is known that there is usually some positive correlation between the total phenolic compounds content and the corresponding antioxidant activity (Andrade et al., 2017; Guiné et al., 2016).

DRYING OF CORIANDER

Drying, or dehydration, is a process through which most of the water is eliminated from food products as a consequence of evaporation of the water molecules, produced in response to the application of heat. This definition disregards other operations that can also remove water from the food, but in a smaller extension, like for example evaporation and mechanical or membrane separations. When dealing with drying and dehydration on a broad spectrum one frequently is faced with concepts that are used interchangeably but, however, at a more precise level, a distinction comes out, so that dehydration applies when the drying treatments are most extreme. In this way, drying applies to the removal of water from fruits such as grapes, dates, apricots or figs, dehydration is used in the case of foods that in the end have practically no water at all, like tea, spices or powdered products like milk, chocolate or potato flakes (Guiné, 2013, 2015).

Drying has been used for thousands of years as a means to preserve foods, being still at present one of the most important processing technologies aimed at extending the shelf life of food products. This goal is achieved through the reduction of water activity, thus inhibiting microbial growth that otherwise could cause deterioration of foods, as well as minimizing chemical reactions and enzymatic activity, often leading to undesirable changes in the chemical composition of physical properties (Guiné et al., 2014; Mota et al., 2010).

Although the main purpose of drying is no doubt the preservation of food, this operation can also be used in combination with complementary processes with specific purposes, like for example in bread baking, during which the heat causes, along with water loss, an expansion of gases and modification in the structure of proteins. Besides, drying also reduces the weight and volume of food, allowing to greatly reduce costs associated with transportation and storage, since the expensive cold storage systems are avoided. However, there are also some situations in which dehydration appears as an unwanted phenomenon, like what happens in the storage of frozen meat (Barroca et al., 2013; Guiné and Fernandes, 2006; Henriques et al., 2012a).

The drying operation modifies in some way both the nutritional value of the food as well as the organoleptic properties, even though this can also be seen as an advantage rather than a disadvantage, since the dried foods typically exhibit colours, aromas, flavours and textures much different form the fresh counterparts, allowing to diverse the offer to consumers who considerably appreciate these products (Guiné, 2015; Henriques et al., 2012b; Santos et al., 2013). One primary cause of change in the quality of some dried or dehydrated foods comes from changes in their internal structure, owing to the loss of a considerable amount of water and collapse of the void spaces, thus altering the texture. Another important visual characteristic that may be influenced by drying (and particularly at higher temperatures and in the presence of oxygen from the drying air) is the colour, due to changes produced through browning and/or caramelization reactions. Finally, but not least important, also changes at the chemical and nutritional levels occur during drying, particularly important in the case of protein denaturation or loss of activity of bioactive compounds (Guiné, 2013).

Like almost all herbs, coriander is a seasonal and highly perishable plant. Hence, it must be preserved in order to be available for consumption all year round. To preserve herbs for a longer period without considerable deterioration in colour and nutritional value of the fresh foliage, appropriate drying is essential to remove the moisture up to a safe water activity level. The most conventional drying techniques used to dehydrate coriander foliage are natural drying (in shade or in the sun), convective hot air drying, vacuum drying, lyophilization, microwave and hybrid microwave drying.

Solar and Convective Drying

Sun-drying is undoubtedly the cheapest drying technique available. However, its long-lasting and climatic dependence can lead to product loss and energy costs. Furthermore, unsafe products can be obtained due to microbial contamination occurring before, during and after the drying process. To minimize the problems caused by weather conditions and contamination, solar drying systems have been developed to dry foods using the energy of the sun but protecting the products from contamination sources and adverse climatic conditions, such as rain. These systems can function with natural or forced air convection (Demir et al., 2004; Navale et al., 2014).

Thermal processes, in which forced convection of air is used for drying foods, are very versatile and faster, but most of them must be avoided in the case of thermo-sensitive products such as fruits, vegetables and herbs. This is justified because high temperatures of the convective drying air are an important cause for quality loss in dried foods. Moreover, lowering the process temperatures ensures a better quality of dried foods, but the drying time and the associated cost increase drastically (Sagar and Suresh Kumar, 2010).

Sagwan et al. (2011) evaluated the effect of different drying processes (shade, oven drying and solar drying) on the sensory characteristics, proximate composition, β-carotene, ascorbic acid, polyphenol and mineral content of dried coriander leaves. The shade drying was performed at room temperature, the solar drying at 54 °C for 6-8 hours and the oven drying at 50±5 °C for 6-8 hours. The scores for sensory quality indicated that coriander dried leaves powder was in the category 'liked moderately', independently of the drying process. Moisture, protein, crude fibre, fat and ash content of shade, solar and oven-dried coriander leaves powder varied from 2.48-2.73, 11.98-12.73, 6.22-6.45, 0.6-0.94 and 14.98-15.57 per cent, respectively. The β-carotene content of the dried coriander leaves obtained by the three drying methods ranged from 18.87 mg/100g in solar dried coriander leaves powder to 21.98 mg/100g in shade-dried coriander leaves powder, while the ascorbic acid content varied from 49.72 mg/100g in solar-dried to 66.87 mg/100g in shade-dried coriander leaves powders. The polyphenol content of coriander leaves powder prepared using different drying methods was fairly similar, ranging from 10.95 to 11.68 mg/100g. Similarly, non-significant differences were obtained regarding the contents of total calcium, iron and copper in coriander leaves powder dried by different methods. Globally, coriander leaves powder prepared using shade, solar and oven drying methods presented good sensory and nutritional properties. Nevertheless, when compared with the fresh counterpart, it is expected to find some losses. Divya (2012) concluded that, even at a mild temperature of 45°C, the oven drying of coriander foliage resulted in substantial loss of both chlorophylls (65%) and carotenoids (35%), as compared with the fresh coriander.

Ahmed (2001) studied the characteristics and quality of coriander leaves after a dry-blanching method. The drying of blanched leaves was performed in a cabinet dryer at a temperature range of 45-80°C. The obtained results revealed that the best quality of dried leaves in terms of chlorophyll and rehydration capacity was found when coriander leaves were blanched in hot water at 80°C for 3 min and dried at 45°C. Under these conditions, the maximum rehydration capacity of coriander leaves was also obtained.

Kaur et al. (2006) investigated the effect of pre-treatments and drying methods on quality and rehydration of coriander leaves for the following conditions: cabinet tray dryer, open sun drying, forced convection air drying, domestic solar dryer with covered and uncovered trays, mini multi-rack solar dryer, portable farm-type solar dryer. Based on the obtained results, the pre-treatment with dipping for 15 min in a solution (0.1% magnesium chloride, 0.1% sodium bicarbonate and 2.0% potassium metabisulphite) in water at room temperature, followed by the drying in mini multi-rack solar dryer, was the best method for obtaining dried coriander leaves with quality (colour) and rehydration characteristics closest to fresh. In addition, dried coriander leaves presented the high overall sensorial acceptability score.

Shaw et al. (2006) investigated the effect of convective drying (at 50°C and velocity of 1.1 m/s) on colour characteristics of coriander foliage (leaves and stems). The results showed that to reduce the moisture content from 93.3% to approximately 12% (wet basis) the dried coriander samples exhibited a significant colour change. The brightness, L value, increased from 26.83 (fresh) to a range of 30.29-33.36 (dried foliage) and the colour change index values varied from 4.59 to 6.58, depending on the replicas tests.

Vacuum Drying

Vacuum drying is a process in which foods are dried in a reduced pressure environment, which lowers the heat necessary to achieve the moisture evaporation and promotes fast drying. Heat-sensitive foods can, therefore, be dried at low temperatures, when under vacuum. Furthermore, this method offers opportunities to shorten the drying time, while the absence of air during dehydration diminishes oxidation reactions. Because of these advantages, the vacuum drying improves the colour, flavour and texture of dried products, when compared to other conventional drying methods (Babu et al., 2018; Nawirska et al., 2009).

Thirugnanasambandham and Sivakumar (2016) evaluated and optimized three key parameters, namely temperature (50-80 °C), loading rate ((0.25–75 kg/m^2), and vacuum (20 to 60 mmHg) to achieve maximum moisture removal from coriander leaves using vacuum drying process. The obtained results revealed that all the process variables affected significantly the drying process, and the optimal conditions of temperature, loading rate and vacuum were, respectively, 75 °C, 0.63 kg/m^2 and 28 mmHg. Under these conditions, it was removed 95% of the moisture initially present and the dried leaves presented 527 mg/100 g of vitamin C content and 13 g/100 g of total dietary fibre.

Freeze Drying and Supercritical CO_2 Drying

Freeze drying is accomplished by controlling the product temperature so that most of the product moisture is removed from a solid-state. Freeze-drying works by freezing the material, then reducing the pressure and adding heat to allow the frozen water in the material to sublimate. In general, since this drying process has no thermal degradation and does not allow the foods' degradation by enzymes, the retention of organoleptic properties, nutritive value and active ingredients are maximized. However, the main drawbacks of freeze-drying are the considerable costs and the low production rate (Babu et al., 2018; Chan et al., 2009).

Carbon dioxide ($scCO_2$) supercritical drying consists of the extraction of water from the product using $scCO_2$. Because the water is not removed by vaporization or sublimation but is instead dissolved in the $scCO_2$, there are no thermal effects on the product quality (Babu et al., 2018). Supercritical carbon dioxide is considered an emerging technology that could assure both the extraction of water as well as a sufficient microbial inactivation. Furthermore, it is showing promising results on the retention of the original structure and quality attributes of foods (Sagar and Suresh Kumar, 2010).

Bourdoux et al. (2018) examined the applicability of supercritical CO2 and freeze-drying to obtain safely dried coriander leaves and stems. Thus, coriander, either fresh or inoculated with three strains of *Escherichia coli* O157:H7, *Salmonella* or *Listeria monocytogenes*, was treated with both drying techniques mentioned. The fresh coriander is a product with a high initial contamination level and a water activity of 0.99. After freeze-drying and $scCO_2$ drying the water, activity was reduced to 0.20±0.04 and 0.21±0.02, respectively. The obtained results revealed that the freeze-drying had a milder effect on inactivation of the natural microbiota when compared to the $scCO_2$, except for mesophilic spores, on which both technologies were ineffective. The results also emphasized that, although freeze-drying brings the highest quality in relation to the organoleptic properties, a high number of spoilers may still be present on the product after dehydration and thus $scCO_2$ drying could be a potential alternative to freeze-drying to provide dried coriander with a reduced number of pathogenic and spoilage microorganisms, particularly yeast and moulds.

Michelino et al. (2018) explored the use of $scCO_2$ drying in combination with High Power Ultrasound (HPU), at different powers (10, 40 and 80 W), different drying times (up to 90 min) and two process temperatures (40 and 50°C), to enhance both the dehydration efficiency and the microbial inactivation on coriander leaves. At the most effective set of experimental conditions tested (40 W; 10 MPa; 40°C), mesophilic bacteria were reduced up to 4 Log, mesophilic spores up to 1 Log, while yeast and moulds were never detected. These two drying processes, $scCO_2$ and HPU + $scCO_2$ (90 min, 40°C, 10 MPa), induced significant total colour differences between the fresh and dried samples, but the modifications in colour were similar in both drying methods tested.

Microwave Drying

Microwave and hybrid microwave (microwave-hot air, microwave-freeze, microwave-vacuum, etc.) drying have become promising techniques of preservation since they supply uniform energy and high thermal conductivity towards the inner surfaces of materials, allied to energy-saving while preventing from any thermal degradation. In addition, microwave drying is a super-fast drying method that can reduce energy costs (Dwivedy et al., 2013). Shorter heating time leads to greater nutrient retention, better quality characteristics, such as texture, colour and flavour, as well as increased production (Alibas, 2009; Di Cesare et al., 2003; Gasmalla et al., 2014).

Sarimeseli (2011) studied the drying of coriander leaves with a microwave drier within the range of power from 180 to 360 W. The results revealed that fresh and dried samples presented no significant differences in the colour parameters and, furthermore, the changes were independent on the power outputs of the microwave drier. Also, the rehydration capacity of the dried leaves decreased as the microwave power increased.

Shaw et al. (2006) concluded that the microwave drying process at a power of 295 W had a small impact on the colour of coriander foliage and the colour change index values for the dried samples ranged from 2.67 to 3.27, corresponding to differences possible to recognize by an experienced observer.

Divya et al. (2012) investigated various power levels (180-850 W) and drying times (30-150 sec) for the microwave drying of coriander. They concluded that there was an increase both in the yield of chlorophylls a and b, as well as on the total carotenoids of coriander foliage at all power levels up to 90 s. As the treatment time increased, carotenoids content decreased, particularly when exposed to lower power levels for longer periods. The analyses of total carotenoids in microwave-dried foliage by HPLC revealed that trans-β-carotene was found to be more stable when compared to other carotenoids, but a partial degradation of lutein and other carotenoids was observed.

Furthermore, the comparison of the profile of carotenoids and quantification of trans-β-carotene revealed that no trans to cis isomerization of β-carotene occurred during the microwave drying for a short time at high power. In addition, the loss of pigments was negligible, allowing obtaining active trans-β-carotene-rich dried foliage for direct use in processed foods. The study also suggested that microwave drying is an efficient cost-effective method for the drying of trans-β-carotene-rich leafy materials.

Sangwan et al. (2011) investigated the effect of microwave drying, with 3 to 4 minutes duration and at the power of 800 W, on the organoleptic acceptability, proximate composition, β-carotene, ascorbic acid and polyphenol content of coriander leaves powder. The leaves summited to microwave drying obtained overall acceptability of 7.40±0.08 (in a 9- point hedonic scale). The moisture content, protein, crude fibre, fat and ash were, respectively, 2.42±0.05, 12.62±0.11, 6.40±0.08, 0.83±0.04 and 15.56±0.15 per cent, in dry matter basis. Furthermore, β-carotene, ascorbic acid and polyphenol content of coriander leaves powder were, respectively, 21.10±01.12, 65.57±1.03 and 11.72±0.11 mg/100g (dry matter basis).

Pati et al. (2015) studied the microwave dehydration of coriander and its effect on quality. The coriander leaves dried at 125 W for 30 minutes possessed significantly higher calcium (282.63 mg/100g dry matter), protein (17.56 g/100 g dm), chlorophyll (3.66 mg/g) and brightness (L= 52.43) when compared to coriander dried at higher output powers, up to 675 W. The results also showed that the shelf life of coriander leaves dried in microwave oven could be predicted to be minimum of 205 days, if stored in metalized polyester (MP) at 65% RH and 30°C temperature.

CONCLUSION

Processing has a great impact on the quality of vegetables and fruits as compared with fresh counterparts. However, having in mind the advantages of processing so as to extend shelf-life or diverse the type of products consumed, some loss of quality is acceptable. Modern technologies and processing operations have emerged in the past years to minimize loss of quality indicators, either at the chemical and nutritional levels or also in terms of organoleptic acceptability, texture, colour and shelf life extension.

ACKNOWLEDGMENTS

The authors would like to acknowledge CI&DETS and CERNAS Research centres from the Polytechnic Institute of Viseu, Portugal.

REFERENCES

Ahmed, J., Shivhare, U. S., and Singh, G. (2001). Drying Characteristics and Product Quality of Coriander Leaves. *Food and Bioproducts Processing* 79, 103–106.

Alibas, I. (2009). Microwave, vacuum, and air drying characteristics of collard leaves. *Drying Technology* 27, 1266–1273.

Andrade, S. C., Guiné, R. P. F., and Gonçalves, F. J. A. (2017). Evaluation of phenolic compounds, antioxidant activity and bioaccessibility in white crowberry (Corema album). *Food Measure 11*, 1936–1946.

Babu, A. K., Kumaresan, G., Raj, V. A. A., and Velraj, R. (2018). Review of leaf drying: Mechanism and influencing parameters, drying methods, nutrient preservation, and mathematical models. *Renewable and Sustainable Energy Reviews 90*, 536–556.

Barroca, M. J., Guiné, R., Alves, M., Oliveira, S., Gonçalves, F., and Correia, P. (2013). Effect of drying on the properties of pears cv. D. Joaquina. In *Proceedings of the VII Congreso Ibérico de Agroingeniería y Ciencias Hortícolas,* (Madrid, Espanha), p. Ref. C0115, 6pp.

Beyzi, E., Karaman, K., Gunes, A., and Buyukkilic Beyzi, S. (2017). Change in some biochemical and bioactive properties and essential oil composition of coriander seed (Coriandrum sativum L.) varieties from Turkey. *Industrial Crops and Products 109*, 74–78.

Bhat, S., Kaushal, P., Kaur, M., and Sharma, H. K. (2014). Coriander (Coriandrum sativum L.): processing, nutritional and functional aspects. *African Journal of Plant Science 8*, 25–33.

Bourdoux, S., Rajkovic, A., De Sutter, S., Vermeulen, A., Spilimbergo, S., Zambon, A., Hofland, G., Uyttendaele, M., and Devlieghere, F. (2018). Inactivation of Salmonella, Listeria monocytogenes and Escherichia coli O157:H7 inoculated on coriander by freeze-drying and supercritical CO_2 drying. *Innovative Food Science & Emerging Technologies 47*, 180–186.

Chan, E. W. C., Lim, Y. Y., Wong, S. K., Lim, K. K., Tan, S. P., Lianto, F. S., and Yong, M. Y. (2009). Effects of different drying methods on the antioxidant properties of leaves and tea of ginger species. *Food Chemistry 113*, 166–172.

Das, U. N. (2018). Arachidonic acid and other unsaturated fatty acids and some of their metabolites function as endogenous antimicrobial molecules: A review. *Journal of Advanced Research 11*, 57–66.

Demir, V., Gunhan, T., Yagcioglu, A. K., and Degirmencioglu, A. (2004). Mathematical Modelling and the Determination of Some Quality

Parameters of Air-dried Bay Leaves. *Biosystems Engineering 88*, 325–335.

Di Cesare, L. F., Forni, E., Viscardi, D., and Nani, R. C. (2003). Changes in the chemical composition of basil caused by different drying procedures. *Journal of Agricultural and Food Chemistry 51*, 3575–3581.

Divya, P., Puthusseri, B., and Neelwarne, B. (2012). Carotenoid content, its stability during drying and the antioxidant activity of commercial coriander (Coriandrum sativum L.) varieties. *Food Research International 45*, 342–350.

Dwivedy, S., Rayaguru, K., and Sahoo, G. R. (2013). Mathematical modeling and quality characteristics of microwave dried medicinal borage leaves. *International Food Research Journal 20*, 769–774.

El-Zaeddi, H., Calín-Sánchez, Á., Nowicka, P., Martínez-Tomé, J., Noguera-Artiaga, L., Burló, F., Wojdyło, A., and Carbonell-Barrachina, Á. A. (2017). Preharvest treatments with malic, oxalic, and acetylsalicylic acids affect the phenolic composition and antioxidant capacity of coriander, dill and parsley. *Food Chemistry 226*, 179–186.

FAOSTAT (2017). *FAOSTAT - Food and agriculture data.*

Gasmalla, M. A. A., Yang, R., Amadou, I., and Hua, X. (2014). Nutritional composition of stevia rebaudiana bertoni leaf: Effect of drying method. *Tropical Journal of Pharmaceutical Research 13*, 61–65.

Guiné, R. P. F. (2013). *Unit Operations for the Food Industry. Volume I: Thermal Processing & Nonconventional Technologies* (Germany: LAP Lambert Academic Publishing).

Guiné, R. P. F. (2015). *Food Drying and Dehydration: Technology and Effect on Food Properties* (Germany: LAP Lambert Academic Publishing GmbH & Co).

Guiné, R., and Gonçalves, F. (2016). Bioactive Compounds in Some Culinary Aromatic Herbs and Their Effects on Human Health. *Mini-Reviews in Medicinal Chemistry 16*, 855–866.

Guiné, R. P. F., and Fernandes, R. M. C. (2006). Analysis of the drying kinetics of chestnuts. *Journal of Food Engineering 76*, 460–467.

Guiné, R. P. F., Cruz, A. C., and Mendes, M. (2014). Convective Drying of Apples: Kinetic Study, Evaluation of Mass Transfer Properties and Data Analysis using Artificial Neural Networks. *Ijfe 10*, 281–299.

Guiné, R. P. F., Gonçalves, C., Gonçalves, F., and Costa, D. (2016). Some Factors that May Affect the Physical-Chemical Properties of Blueberries. Agricultural Engineering International: *CIGR Journal 18*, 334–342.

Henriques, F., Guiné, R., and Barroca, M. J. (2012a). Chemical properties of pumpkin dried by different methods. *Croatian Journal of Food Technology, Biotechnology and Nutrition 7*, 98–105.

Henriques, F., Guiné, R. P. F., and Barroca, M. J. (2012b). Influence of Drying Treatment on Physical Properties of Pumpkin. *Croatian Journal of Food Technology, Biotechnology and Nutrition 7*, 53–58.

Kasonga, A. E., Kruger, M. C., and Coetzee, M. (2019). Free fatty acid receptor 4-β-arrestin 2 pathway mediates the effects of different classes of unsaturated fatty acids in osteoclasts and osteoblasts. *Biochimica et Biophysica Acta (BBA) - Molecular and Cell Biology of Lipids 1864*, 281–289.

Kaur, P., Kumar, A., Arora, S., and Singh Ghuman, B. (2006). Quality of dried coriander leaves as affected by pretreatments and method of drying. *European Food Research and Technology 223*, 189–194.

Khodadadi, M., Dehghani, H., Jalali Javaran, M., and Christopher, J. T. (2016). Fruit yield, fatty and essential oils content genetics in coriander. *Industrial Crops and Products 94*, 72–81.

Kozłowska, M., Gruczyńska, E., Ścibisz, I., and Rudzińska, M. (2016). Fatty acids and sterols composition, and antioxidant activity of oils extracted from plant seeds. *Food Chemistry 213*, 450–456.

Lichtenstein, A. H., Appel, L. J., Brands, M., Carnethon, M., Daniels, S., Franch, H. A., Franklin, B., Kris-Etherton, P., Harris, W. S., Howard, B., et al. (2006). Diet and lifestyle recommendations revision 2006: a scientific statement from the American Heart Association Nutrition Committee. *Circulation 114*, 82–96.

Mandal, S., and Mandal, M. (2015). Coriander (Coriandrum sativum L.) essential oil: Chemistry and biological activity. *Asian Pacific Journal of Tropical Biomedicine 5*, 421–428.

Michelino, F., Zambon, A., Vizzotto, M. T., Cozzi, S., and Spilimbergo, S. (2018). High power ultrasound combined with supercritical carbon dioxide for the drying and microbial inactivation of coriander. *Journal of CO_2 Utilization 24*, 516–521.

Mohamed, S. (2014). Functional foods against metabolic syndrome (obesity, diabetes, hypertension and dyslipidemia) and cardiovasular disease. *Trends in Food Science & Technology 35*, 114–128.

Momin, A. H., Acharya, S. S., and Gajjar, A. V. (2012). Coriandrum sativum—review of advances in phytopharmacology. *International Journal of Pharmaceutical Sciences and Research 3*, 1233–1239.

Mota, C.L., Luciano, C., Dias, A., Barroca, M. J., and Guiné, R. P. F. (2010). Convective drying of onion: Kinetics and nutritional evaluation. *Food and Bioproducts Processing 88*, 115–123.

Nakamura, H., Tsujiguchi, H., Kambayashi, Y., Hara, A., Miyagi, S., Yamada, Y., Nguyen, T. T. T., Shimizu, Y., Hori, D., and Nakamura, H. (2019). Relationship between saturated fatty acid intake and hypertension and oxidative stress. *Nutrition 61*, 8–15.

Navale, S. R., Supriya, U., Harpale, V. M., and Mohite, K. C. (2014). Effect of Solar Drying on the Nutritive Value of Fenugreek Leaves. *International Journal of Engineering and Advanced Technology 4*, 4133–4136.

Nawirska, A., Figiel, A., Kucharska, A. Z., Sokół-Łętowska, A., and Biesiada, A. (2009). Drying kinetics and quality parameters of pumpkin slices dehydrated using different methods. *Journal of Food Engineering 94*, 14–20.

Pati, G. D., Pardeshi, I. L., and Shinde, K. J. (2015). Drying of green leafy vegetables using microwave oven dryer. *Journal of Ready to Eat Food 2*, 18–26.

Praagman, J., Vissers, L. E. T., Mulligan, A. A., Laursen, A. S. D., Beulens, J. W. J., van der Schouw, Y. T., Wareham, N. J., Hansen, C. P., Khaw, K. T., Jakobsen, M. U., et al. (2019). Consumption of individual saturated fatty acids and the risk of myocardial infarction in a UK and a Danish cohort. *International Journal of Cardiology 279*, 18–26.

Sagar, V. R., and Suresh Kumar, P. (2010). Recent advances in drying and dehydration of fruits and vegetables: a review. *Journal of Food Science and Technology 47*, 15–26.

Sahib, N. G., Anwar, F., Gilani, A. H., Hamid, A. A., Saari, N., and Alkharfy, K. M. (2013). Coriander (Coriandrum sativum L.): A Potential Source of High-Value Components for Functional Foods and Nutraceuticals- A Review. *Phytotherapy Research 27*, 1439–1456.

Sangwan, A., Kawatra, A., and Sehgal, S. (2011a). Bio-chemical analysis of coriander leaves powder prepared using various drying methods. *Journal of Dairying Foods & Home Sciences 30*, 202–205.

Sangwan, A., Kawatra, A., and Sehgal, S. (2011b). *Bio-chemical analysis of coriander leaves powder prepared using various drying methods.*

Santos, S. C. R. V. L., Guiné, R. P. F., and Barros, A. I. A. (2013). Influence of Drying on the Properties of Pears of the Rocha Variety (Pyrus communis. *International Journal of Food Engineering 9*.

Sarimeseli, A. (2011). Microwave drying characteristics of coriander (Coriandrum sativum L.) leaves. *Energy Conversion and Management 52*, 1449–1453.

Sayed-Ahmad, B., Talou, T., Saad, Z., Hijazi, A., and Merah, O. (2017). The Apiaceae: Ethnomedicinal family as source for industrial uses. *Industrial Crops and Products 109*, 661–671.

Schtscherbyna, A., Ribeiro, B.G., and Maria Lucia Fleiuss, F. (2019). Chapter 6 - Bone Health, Bone Mineral Density, and Sports Performance. In *Nutrition and Enhanced Sports Performance* (Second Edition), D. Bagchi, S. Nair, and C.K. Sen, eds. (Academic Press), pp. 73–81.

Shahidi, F. (2005). *Bailey's Industrial Oil and Fat Products* (New York, NY, USA: Wiley-Interscience).

Shaw, M., Meda, V., Tabil, L., and Opoku, A. (2006). Drying and color characteristics of coriander foliage using convective thin-layer and microwave drying. *Journal of Microwave Power and Electromagnetic Energy 41*, 56–65.

Sun, B., Luo, C., Cui, W., Sun, J., and He, Z. (2017). Chemotherapy Agent-unsaturated Fatty Acid Prodrugs and Prodrug-Nanoplatforms for Cancer Chemotherapy. *Journal of Controlled Release 264*, 145–159.

Sun, C., Wang, R., Li, Z., and Zhang, D. (2019). Dietary magnesium intake and risk of depression. *Journal of Affective Disorders 246*, 627–632.

Szempliński, W., Nowak, J., and Jankowski, K. J. (2018). Coriander (*Coriandrum sativum* L.) Response to Different Levels of Agronomic Factors in Poland. *Industrial Crops and Products 122*, 456–464.

Thirugnanasambandham, K., and Sivakumar, V. (2016). Enhancement of shelf life of Coriandrum sativum leaves using vacuum drying process: modeling and optimization. *Journal of the Saudi Society of Agricultural Sciences 15*, 195–201.

Uitterhaegen, E., Sampaio, K. A., Delbeke, E. I. P., De Greyt, W., Cerny, M., Evon, P., Merah, O., Talou, T., and Stevens, C. V. (2016). Characterization of French Coriander Oil as Source of Petroselinic Acid. *Molecules 21*, 1202.

USDA (2018). *Food Composition Databases*.

Vu, K. D., Carlettini, H., Bouvet, J., Côté, J., Doyon, G., Sylvain, J. F., and Lacroix, M. (2012). Effect of Different Cranberry Extracts and Juices During Cranberry Juice Processing on the Antiproliferative Activity Against Two Colon Cancer Cell Lines. *Food Chemistry 132*, 959–967.

Wangensteen, H., Samuelsen, A. B., and Malterud, K. E. (2004). Antioxidant activity in extracts from coriander. *Food Chemistry 88*, 293–297.

Zhang, X., Shen, Y., Prinyawiwatkul, W., King, J. M., and Xu, Z. (2013). Comparison of the activities of hydrophilic anthocyanins and lipophilic tocols in black rice bran against lipid oxidation. *Food Chemistry 141*, 111–116.

In: Coriander
Editor: Deepak Kumar Semwal

ISBN: 978-1-53616-483-1
© 2019 Nova Science Publishers, Inc.

Chapter 3

ANTIOXIDANT AND ANTIMICROBIAL ACTIVITIES OF CORIANDER (*CORIANDRUM SATIVUM*)

Miroslava Kačániová[1,2,]* and Eva Ivanišová[3]

[1]Slovak University of Agriculture, Nitra, Slovakia
[2]Department of Bioenergy and Food Technology, University of Rzeszow, Poland
[3]Department of Plant Storage and Processing, Slovak University of Agriculture, Nitra, Slovakia

ABSTRACT

Coriander (*Coriandrum sativum* L.), a member of the *Apiaceae* family, is among the most widely used medicinal plant, possessing nutritional as well as medicinal properties. All parts of this plant are edible, although they have very distinct flavours and uses. The various compound extracts and essential oil from coriander have been proven to possess antibacterial, anti-oxidant, free radical, antidiabetic, anticancerous,

---

*Corresponding Author's Email: miroslava.kacaniova@gmail.com.

antimutagenic and other activities. Antioxidant activities of *C. sativum* reported in the literature mostly focused on the aerial parts of the herb. The nature of the extracting solvent is one of the most important factors in the extraction of antioxidants and bioactive compounds. The ethyl acetate extracts of *C. sativum* leaf and seed have been reported with the highest phenolic content. In a different study, the aqueous extract of *C. sativum* leaf and shoots exhibited antioxidant activity in a β-carotene/linoleic acid model. The composition of phenolic fraction present in coriander is still incompletely studied and some data are contradictory. Phytochemicals derived from coriander leaves were found to have antibacterial activity against *Salmonella choleraesuis*, and this activity was found to be caused in part by these chemicals acting as nonionic surfactants. The essential oil and its fractions could be used as potential antimicrobial agents to treat or prevent *Candida* yeast infections. Antimicrobial potential of aqueous infusions and aqueous decoctions of *Coriandrum sativum* against 186 bacterial isolates belonging to 10 different genera of Gram-positive bacterial population and 2 isolates of *Candida albicans* isolated from urine specimens. Essential oils from *Coriandrum sativum* was evaluated against *Candida* spp. isolates from the oral cavity of patients with periodontal disease. The extracted oil was screened for antimicrobial activity against both Gram-positive (*Staphylococcus aureus*, *Bacillus* spp.) and Gram-negative (*Escherichia coli*, *Salmonella typhi*, *Klebsiella pneumonia*, *Proteus mirabilis*, *Pseudomonas aeruginosae*) bacteria and a pathogenic fungus, *Candida albicans*. Only *P. aeruginosae* showed resistance to CEO, while other tested bacteria were highly affected. In the food industry, the quality and safety of prepared or processed foods are of prime importance. The microorganisms present in food can lead to spoilage and deterioration of the quality of food products, and if ingested by humans can cause infection and illness.

Keywords: herbals, biological activity, foods, human body, essential oils

INTRODUCTION

Coriander (*Coriandrum sativum* L.) is an annual herbaceous plant (Figure 1) belongs to the *Apiaceae* family (Gupta 2019). It is originated from the Mediterranean area but is extensively cultivated in North Africa, Central Europe and Asia as a culinary and medicinal herb. In addition, to India, it is also cultivated in Romania, France, Spain, Furthermore, it is successfully grown in a wide range of conditions (Laribi et al., 2015). The plant is strongly aromatic, erect, herbaceous about 1.5 m height with a hollow stem. It has shiny, bright green leaves, umbels of small white or pale-pink flowers. The leaves and stems are called cilantro, while the fruits are called coriander. The fruits (seeds) are globular, uniform light-brown, round consisting of two pericarps with a warm pleasant odour (Charles, 2013; Al-Snafi, 2016).

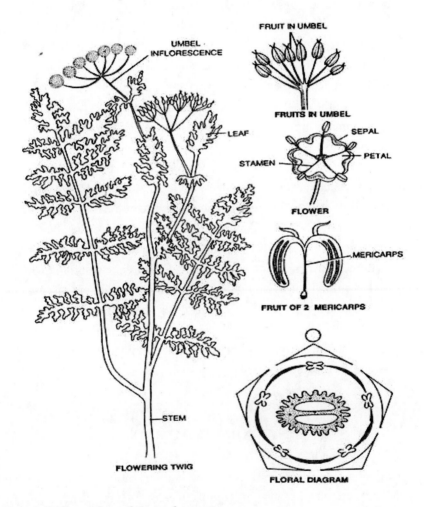

Figure 1. Different parts of *Coriandrum sativum*.

The fruits (seeds) are sweet, spicy; the leaves have a very characteristic and distinctive aroma (cilantro). The seeds are mild, warm and sweet, spicy, fruity, with a slightly citrusy and minty undertone. Immature fruits and leaves have an unpleasant odour called a "stink bug smell" which is due to the presence of *trans*-tridecen in the oil. On the other hand, the fragrance in the mature fruits pleasantly is similar to citrus peel and sage The essential oil flavour is warm, spicy-aromatic, sweet, and fruity (Charles, 2013; Önder, 2018).

Coriander closely resembles flat leaf parsley. This resemblance makes many people confused between the two however, coriander has a strong fragrance and parsley has a mild fragrance. It grows best in dry climates however it can grow in any type of soil like light, well aired, moist, loamy soil, and light to heavy black soil (Verma et al., 2011).

All parts of the plant are edible, fresh leaves can be used for garnishing and are a common ingredient in many foods like chutneys and salads. The green herb is also employed for the preparation of either steam-distilled essential oil or the solvent extracted oleoresin. Fresh juice of coriander is extremely advantageous in curing many deficiencies related to vitamins and iron. One to two teaspoons of its juice added to refreshing buttermilk is incredibly beneficial in curing many diseases. Fresh leaves can be eaten as such because of various health benefits however if it is not harvested freshly seeds mature and ripen in late summer developing delicate aroma which is then used as a dried spice. Moreover, this plant is used to cure diseases like digestive tract disorders, respiratory tract disorders, urinary tract infections. Coriander has been reported to possess many pharmacological activities like antioxidant, antidiabetic, antimutagenic, antilipidemic, antispasmodic (Charles, 2013; Bhat et al., 2014).

ANTIOXIDANT ACTIVITY

Coriander has been shown to have strong antioxidant activity and medicinal properties because of its bioactive constituents. Ripe, dried fruit (Table 1) contains moisture 11%, 23-36% fibre, 13-20% carbohydrates, 16-28% fatty oil, 11-17% proteins, 5% mineral compounds and essential oil in an amount of 1-3% (Charles, 2013; Chizoba, 2015). The major constituent of the oil is D-linalool (55-90%), neryl acetate, γ-terpinene, camphor, α-

pinene, and geranyl acetate. Yildiz (2016) identified by gas chromatography/ mass spectrometry analysis 19 compounds representing 95.30% of the essential oil: (E)-2-decenal (29.87%), linalool (21.61%), (E)-2-dodecenal (7.03%), dodecanal (5.78%), (E)-2-undecenal (3.84%), (E)-2-tridecenal (3.56%), (E)-2-hexadecenal (2.47%), tetradecenal (2.35%), and α-pinene (1.64%). The major compounds in the fruit (seeds) and plants are tocopherols, carotenoids and chlorophylls and sugars, ascorbic acid, phenolic, flavonols and anthocyanins (Table 2) (Charles, 2013; Nadeem et al. 2013; Bhat, 2014; Al-Snafi, 2016).

Table 1. The nutrient compounds of dried coriander fruits

Nutrient	Units
Water	8.86 g/100 g
Energy	298 kcal
Protein	12.37 g/100 g
Total lipid	17.77 g/100 g
Total dietary fibre	41,9 g/100 g
Carbohydrate, by difference	54.99 g/100 g
Calcium	709 mg/100 g
Magnesium	330 mg/100 g
Phosphorus	409 mg/100 g
Zinc	4.7 mg/100 g
Iron	16.32 mg/100 g
Potassium	1267 mg/100 g
Sodium	35 mg/100 g
Vitamin C	21 mg/100 g
Fatty acids, total saturated	0.990 g/100 g
Fatty acids, total monosaturated	13.580 g/100 g
Fatty acid, total polyunsaturated	1.750 g/100 g

The ethanolic extract (300 mg/kg) of *C. sativum* was shown to possess hepatoprotective activity which may be due to the antioxidant potential of phenolic compounds (Pandey et al., 2011). In the study of Yildiz (2016), the samples *C. sativum* were screened for their antioxidant activities using 2,2-diphenyl-1-picrylhydrazyl radical scavenging and β-carotene bleaching assay. The IC_{50} value for ethanol extract (20–100 μg/mL) of *C. sativum* was determined as 74.87 ± 0.03 µg/mL. Total antioxidant activity value for *C. sativum* ethanol extract was 85.85 ± 0.04%. Total phenolic content for ethanol extract of the plant was determined as 14.97 ± 0.05 mg gallic acid equivalents/g dry weight. The fruits of *C. sativum* showed scavenging activity against superoxides and hydroxyl radicals and inhibited the process of peroxidative damage in diabetic rats (10 g/100 g feeds) It was also reactivated the antioxidant enzymes and antioxidant levels in diabetic rats (Deepa and Anuradha, 2011). Wu et al. (2010) demonstrated that the aerial parts of coriander had a strong anti-inflammatory property which inhibits proinflammatory mediator expression by suppressing NF-kappaB activation and MAPK signal transduction pathway in LPS-induced macrophages. The administration of coriander significantly protected against lead-induced oxidative stress in mice testis (Sharma et al., 2010).

Table 2. The nutrient compounds and antioxidant activity of raw coriander leaves

Nutrient	Units
Water	92.21 g/100 g
Energy	23 kcal
Protein	2.13 g/100 g
Total lipid	0,52 g/100 g
Ash	1.47 g/100 g
Total dietary fibre	2,8 g/100 g
Carbohydrate, by difference	3,67 g/100 g
Calcium	67 mg/100 g
Phosphorus	60 mg/100 g
Iron	10 mg/100 g
Vitamin C	27 mg/100 g
Vitamin B_6	0,149 mg/100 g
Vitamin E	2,5 mg/100 g
Antioxidant activity (ORAC)	5,141 µmol/100 g
Total polyphenol content	151 mg GAE/100 g

GAE – gallic acid equivalent.

The ethyl acetate extract of *C. sativum* roots showed the highest antiproliferative activity on MCF-7 cells (IC_{50} = 200.0 ± 2.6 µg/mL) and had the highest phenolic content, FRAP and DPPH scavenging activities among the extracts. *C. sativum* root (0–500 µg/ml) inhibited DNA damage and prevented MCF-7 cell migration induced by H_2O_2, suggesting its potential in cancer prevention and inhibition of metastasis. The extract exhibited anticancer activity in MCF-7 cells by affecting antioxidant enzymes possibly leading to H_2O_2 accumulation, cell cycle arrest at the G2/M phase and apoptotic cell death by the death receptor and mitochondrial apoptotic pathways (Tang et al. 2013, 2). In the study of Msaada et al. (2017) fruit methanolic extract of three coriander (*Coriandrum sativum* L.) varieties (Tunisian, Syrian and Egyptian) was assayed for their antioxidant activities. The RP-HPLC analysis revealed the identification of phenolics in coriander fruits with chlorogenic and gallic acids as main compounds in Tunisian, Syrian and Egyptian varieties, respectively. Moreover, fruit methanolic extracts exhibited remarkable DPPH radical scavenging activity with IC_{50} values ranged from 27.00 ± 6.57 to 36.00 ± 3.22 µg/mL. EC_{50} values of reducing power activity varied significantly ($P < 0.05$) from 54.20 ± 6.22 to 122.01 ± 13.25 µg/mL. The IC_{50} values of β-carotene bleaching assay were between 160.00 ± 18.63 and 240.00 ± 26.35 µg/mL. Linalool (0.4 µM and 2 µM) obtained from coriander was found to decrease cell viability of HepG2 cells, inhibit complexes I and II and decrease adenosine triphosphate (ATP). It is also increased reactive oxygen species generation and decreased glutathione (Usta et al., 2009).

The study of Darughe et al. (2012) observed the effect of essential oil from coriander (CEO) in the cake as a natural antioxidant and proposed as potential substitutes of synthetic antioxidants in food preservation. Then, antioxidant and antifungal activities of the CEO were evaluated in the cake during 60-day storage at room temperature. The results indicated that CEO at 0.05, 0.10 and 0.15% inhibited the rate of primary and secondary oxidation products formation in cake and their effects were almost equal to BHA at 0.02% ($p < 0.01$). In a study of Marangoni and Moura (2011, 124) four formulations of Italian salami type were produced: without antioxidants; with essential oil of coriander essential oil (0.01%); with BHT

(0.01%); and with coriander essential oil and BHT (0.005 and 0.005%). The antioxidant activity of salamis was evaluated by the lipid oxidation, through the techniques of peroxide number and TBARS. The salami with the coriander essential oil exhibited a reduction in lipid oxidation by increasing the shelf life of the product. The salami with the coriander essential oil and BHT showed no synergism between the antioxidants. The salami using BHT presented less antioxidant activity than that of the salami using coriander essential oil. Treatment of heart failure rats with *C. sativum* orally (1g/kg b.w.) improved the altered hemodynamics, restored the cardiac antioxidant enzymes armory, attenuated oxidative stress, improved lipid profile, lowered atherogenic indices, decreased the levels of ETA and ETB receptor mRNA and protein, and restored the cardiac morphology (Siddiqi et al. 2017, 119). In a study of Nithya and Sumalatha (2014) the efficacy of *Coriandrum sativum* against HT-29 cell lines showed that the incubation of cancer cells reduced the viability of HT-29 cell lines and the dead cells were significantly increased with high extract concentration. Hence ethanolic extract (10 mg/ml) of *Coriandrum sativum* exhibited high cytotoxicity.

The work of Saeed et al. (2014) has been carried out in order to investigate the possible ameliorative effect of coriander seeds methanolic extract (CSE) on sodium arsenite (As)-induced toxicity in albino rats. 2,3 Dimercaptosuccinic acid (DMSA) was used as a standard chelating agent. The experiment lasted for 8 weeks and blood samples were withdrawn 4 and 8 weeks after as and other treatments administrations. As was used to induce hepatotoxicity in rats in a dose equivalent to 100 p.p.m. *In-vivo* studies using the different biochemical techniques employed in hepato-renal functions as well as liver apoptotic DNA and total RNA alterations proved that as caused a significant increase in parameters concerned to hepato-renal toxicity while treatment of CSE or DMSA caused an ameliorative effect on this toxicity. Administration of CSE and DMSA together along with arsenite proves the synergistic effects of these chelating agents on arsenite toxicity. Explore the anti-anxiety activity of hydroalcoholic extract of *Coriandrum sativum* using different animal models (elevated plus maze, open field test, light and dark test and social interaction test) of anxiety in mice was tested in a study of Mahendra and Bisht, (2011). Diazepam (0.5 mg/kg) was used as the standard and dose of hydroalcoholic extract of *C. sativum* fruit (50, 100 and 200 mg/kg) was selected as per OECD guidelines. Results suggested that extract of *C. sativum* at 100 and 200 mg/kg dose produced anti-anxiety effects almost similar to diazepam, and at 50 mg/kg dose did not produce anti-anxiety activity on any of the paradigm used.

Daily administration of *C. sativum* extract (100, 200 and 300 mg/kg body weight) 1 h prior to induction of stress significantly decreased the stress-induced urinary levels of VMA (vanillylmandelic acid) from 382.79 ± 10.70 to 350.66 ± 15.15, 291.21 ± 16.53 and 248.86 ± 13.56 µg/kg/24 h and increased the ascorbic acid excretion levels from 66.73 ± 9.25 to 69.99 ± 7.37, 105.28 ± 13.74 and 135.32 ± 12.54 µg/kg/24 h at 100, 200 and 300 mg/kg, respectively, in a dose-dependent fashion without affecting the normal levels in control groups. The amnesic deficits (acquisition, retention and recovery) induced by scopolamine (1 mg/kg, i.p.) in rats was reversed by *C. sativum* dose-dependently. The extract also inhibited lipid peroxidation in both rat liver and brain to a greater extent than the standard antioxidant, ascorbic acid (Koppula and Choi, 2012). In a study of Hwang et al. (2014) was examined whether coriander ethanol extract (CSE) has protective effects against UVB-induced skin photoaging in normal human dermal fibroblasts (NHDF) *in vitro* and in the skin of hairless mice *in vivo*. The main component of CSE, linolenic acid, was determined by gas chromatography-mass spectroscopy. Authors measured the cellular levels of procollagen type I and MMP-1 using ELISA in NHDF cells after UVB irradiation. NHDF cells that were treated with CSE after UVB irradiation exhibited higher procollagen type I production and lower levels of MMP-1 than untreated cells. In the study was found that the activity of transcription factor activator protein-1 (AP-1) was also inhibited by CSE treatment. It was measured the epidermal thickness, dermal collagen fibre density, and procollagen type I and MMP-1 levels in photo-aged mouse skin *in vivo* using histological staining and western blot analysis. The results showed that CSE-treated mice had thinner epidermal layers and denser dermal collagen fibres than untreated mice.

In a study of Talaei et al. (2017) was twenty-four adult male Wistar rats randomly divided to 4 groups: healthy control, iron overload, iron overload treated with deferoxamine, and 200 mg/kg of *C. sativum* extract, respectively. Assessments were performed for serum iron, ferritin, and serum markers of hepatic and cardiac damages, antioxidant enzyme, and lipid peroxidation levels. The results indicated serum iron and ferritin concentration were significantly increased in the iron overload group compared to healthy controls ($P < 0.05$). Treatment with *C. sativum* extract significantly decreased iron and ferritin concentration ($P < 0.05$). Histopathological changes in the liver, kidney, and the cardiac system, and iron accumulation in the liver were observed in the iron overload group compared to healthy controls ($P < 0.05$). Treatment with *C. sativum* extract significantly decreased biochemical parameters, such as alanine aminotransferase (ALT), aspartate aminotransferase (AST), alkaline phosphatase (ALP), lactate dehydrogenase (LDH) and creatinine phosphokinase (CPK), and improved tissue damage and decreased iron accumulation in the liver compared to the iron overload group ($P < 0.05$).

In vitro, antioxidant, anticancer, cytotoxic and antiviral activities of *C. sativum* seeds were investigated by Fayyad and Yaakobi (2017). To detect the antioxidant activity of methanol and hexane extracts of C. *sativum* seeds, assays to demonstrate its reducing power and total phenolic content was performed by spectrophotometry. The cytotoxic and anticancer effect of methanol, hexane and aqueous extracts was assessed using MTT assay. Additionally, the antiviral activity of methanol, hexane and aqueous extracts against HSV-1 was investigated by plaque reduction assay. Based on absorbance readings, the reducing power of methanol and hexane extracts of *C. sativum* seeds were approximately equal, while the absorbance readings indicated that the total phenol content assay of methanol extract was higher than that of the hexane extract. Based on MTT assay, both hexane and aqueous extracts exhibited inhibition of Vero cells with CC_{50} of 600 μg/ml and 700 μg/ml, respectively, while the minimum inhibition of HepG2 cells was observed at a concentration of 350 μg/ml for the three extracts. The aqueous and hexane extracts exhibited reductions in the formation of HSV-1 plaques with an observed IC_{50} of 350 μg/ml and 250 μg/ml, respectively. Based on their obtained results, the hexane, methanol and aqueous extracts of *C. sativum* seeds demonstrated anti-cancer and antiviral effects while the hexane and methanol extracts demonstrated antioxidant potency.

USAGE AND APPLICATIONS IN FOOD SCIENCE

Antioxidant and Food Preservation Studies have shown that essential oils are better natural antioxidant and may be used as a substitute for the synthetic antioxidants in food preservation. Misharina and Polshkov (2005) showed that the mixture of essential oils from laurel and coriander possessed antioxidant properties and strongly inhibited the oxidation of the components of the fennel oil during various storage conditions. Similarly, Ramadan and Wahdan (2012) prepared blends (10% and 20%, w/w) of black cumin seed oil and coriander seed oil with corn oil. The results of their studies showed that oxidative stability of oil blends was better than its original oil, due to the consequent changes in the fatty acid and tocopherol profiles and minor bioactive lipids found in coriander and black cumin oils. Coriander essential oil containing camphor (44.99%), cyclohexanol acetate, limonene (7.17%), and α-pinene (6.37%) was evaluated in the cake during 60 days' storage at room temperature (Darughe et al., 2012). The results showed that 0.05%, 0.10%, and 0.15% coriander essential oil inhibited the rate of primary and secondary oxidation product formation in the cake and their effects were almost equal to butylated hydroxyl aniline at 0.02%. Organoleptic evaluation of cakes containing 0.05% coriander essential oil was not different from the control. These results showed that the coriander essential oil could be used as natural antioxidants in lipid-containing foodstuffs.

ANTIBACTERIAL APPLICATIONS

In the food industry, the quality and safety of prepared or processed foods are of prime importance. The microorganisms present in food can lead to spoilage and deterioration of the quality of food products, and if ingested by humans can cause infection and illness. Thus food manufacturers try their best to reduce or eliminate microorganisms from food products. It has been estimated that about one-third of the world's food production is lost annually on account of microbial spoilage or contamination (Alboofetileh et al., 2014).

Freeze-dried coriander was extracted by petroleum ether, 95% ethanol, and water, respectively. The antibacterial experiment indicated that only water extract presented significant antimicrobial activity and the minimum inhibition concentration (MIC) was below 10% of the original extract. The inhibition effect of coriander extracts on microorganisms and the effects of pH, temperature and NaCl concentration on its antimicrobial activities were evaluated. The results showed that antibacterial activity of coriander extracts was stable under heating and had the best antibacterial effects at pH 6 with 2.0% NaCl concentration (Cao et al., 2012).

The coriander seed extract in 80% methanol was examined for the polyphenol composition and the antimicrobial potential against pathogenic bacteria *Escherichia coli*, *Pseudomonas aeruginosa*, *Staphylococcus aureus*, and *Bacillus pumilus*. The antimicrobial activity was examined by determining cell damage and measuring cell inhibition zone. The growth inhibition zones observed by agar well diffusion method are 11.93 to 17.27 mm in diameter in presence of coriander extract. The minimum inhibitory concentration of coriander was 4.16 mg/ml for the bacteria tested. Increased release of intracellular nucleotides and proteinaceous materials from the bacterial cells in the presence of methanolic coriander extract suggests that the primary mechanism of action of coriander extract is membrane damage, which leads to cell death. The results obtained herein further encourage the use of coriander in antibacterial formulations due to the fact that polyphenol-rich coriander extract could effectively kill pathogenic bacteria related to foodborne diseases (Dua et al., 2014).

To investigate the antimicrobial activity of ethanol, methanol, acetone, chloroform, hexane and petroleum ether extract of *Coriandrum sativum* was tested against infectious disease causing bacterial pathogens such as *E. coli*, *Pseudomonas aeruginosa*, *Staphylococcus aureus* and *Klebsiella pneumonia*, fungus like *Aspergillus niger, Candida albicans, Candida kefyr* and *Candida tropicalis* using the agar well diffusion method. The methanol extract of *Coriandrum sativum* showed more activity against *Staphylococcus aureus* zone of diameter 12.17 ± 0.29 and *Klebsiella pneumonia* zone of diameter 12.17 ± 0.15 and the methanol extract of *Coriandrum sativum* showed more activity against *Candida albicans* zone of diameter 14.20 ± 0.20 and *Aspergillus niger* of diameter 10.10 ± 0.10 when compared to other solvent extracts. In the present study, both in bacteria and fungi methanol extract showed a varying degree of inhibition to the growth of tested organism than ethanol, acetone, chloroform, hexane and petroleum ether. The results confirmed that the presence of antibacterial and antifungal activity in the sundried extract of *Coriandrum sativum* against the human pathogenic organisms. The methanolic extract of sun-dried *Coriandrum sativum* showed better activity against the most tested organisms (Ratha bai and Kanimozhi, 2012).

Besides, coriander seed extract has an effective antimicrobial activity against Gram-negative (*Klebsiella pneumonia* and *Pseudomonas aeruginosa*) and Gram-positive bacteria (*Staphylococcus aureus* PTCC1431) (Zare-Zardini et al., 2012).

Matasyoh et al. (2009) explained that the antibacterial activity exhibited by the *C. sativum* leaf oil can be attributed to the synergic effect of the antimicrobial agents present in the oil. The leaf oil contains 44 compounds mostly of aromatic acids of which the major is 2-decenoic acid, E-11-tetradecenoic acid, capric acid, undecyl alcohol and tridecanoic acid. The high concentration of 2-decenoic acid in leaf oil makes it potentially useful in medicines and perfumes.

The effects of coriander seed extracts, cefoxitin and extract + cefoxitin on clinically resistant strains (MRSA, *E. coli, K. pneumoniae*) were studied. These pathogen microorganisms were chosen due to their importance in several infections such as urinary tract infection, endocarditis, septicemia, osteomyelitis and hospital-acquired infection and have been reported to be antibiotic resistant. In disc diffusion assay, the ME, PE extracts and cefoxitin inhibition zones of the MRSA strain were identified as 8,10 and 18 mm, respectively. However, the combination of extracts with cefoxitin was increased and inhibition zones were determined as 21 mm for the cefoxitin+1250 μg/ml ME extract and 22 mm for the cefoxitin+1250 μg/ml PE extract. Comparable results were also obtained for the *E. coli* isolate. Both combinations showed an increase in antibacterial activity in all tested microorganisms. However, according to the FIC results, there was no synergistic interaction on MRSA and *E. coli* by the checkerboard method. Cefoxitin showed a small zone of inhibition, but when used with plant extracts (especially with PE extract), it exhibited a larger zone of inhibition against ESBL positive *K. pneumoniae*. The inhibition zones of *K. pneumoniae* were determined as 15 mm for cefoxitin; 9 mm for the ME extract; 25 mm for the PE extract; 25 mm for the cefoxitin+ 1250 μg/ml ME extract and 34 mm for the cefoxitin + 1250 μg/ml PE extract. The susceptibility of *K. pneumoniae* increased with cefoxitin + 1250 μg/ml PE extract and inhibition zone was measured as 34 mm. According to the interaction results of the checkerboard method, *K. pneumoniae* showed synergistic interaction against combinations of cefoxitin ME and PE extracts at 0.03516 and 0.03125 FIC index, respectively (Ildiz et al., 2018).

ANTIMICROBIAL ACTIVITY OF CORIANDER ESSENTIAL OIL

The essential oil derived from coriander plants has been reported to exhibit exceptionally good antimicrobial effects against bacteria, yeasts, fungi, and viruses. CEO and its various fractional distillates were effective antimicrobial agents particularly against *Listeria monocytogenes* because of the presence of long chain (C6–C10) alcohols and aldehydes. Mixing of different fractions showed additive, synergistic, or antagonistic effects against individual test microorganisms (Delaquis et al. 2002). Similarly, Matasyoh et al. (2009) obtained essential oil from the leaves by hydrodistillation evaluated for *in vitro* antimicrobial activity. The oil was dominated by aldehydes and alcohols which accounted for 56.1% and 46.3% of the oil, respectively. The extracted oil was screened for antimicrobial activity against both Gram-positive (*Staphylococcus aureus*, *Bacillus* spp.) and Gram-negative (*Escherichia coli*, *Salmonella typhi*, *Klebsiella pneumonia*, *Proteus mirabilis*, *Pseudomonas aeruginosa*) bacteria and a pathogenic fungus, *Candida albicans*. Only *P. aeruginosa* showed resistance to CEO, while other tested bacteria were highly affected.

Coriander essential oil has been reported to inhibit a broad spectrum of microorganisms (Caseti et al., 2012; Pawar Vinita et al., 2013; Thompson et al., 2013) and has proven its efficacy as an antibacterial agent (Cantore et al., 2004; Kubo et al., 2004; Vejdani et al., 2006; Deb Roy et al., 2010; Duman et al., 2010; Lixandru et al., 2010; Rattanachaikunsopon and Phumkhachorn, 2010; Silva et al., 2011; Yadav et al., 2012).

Rattanachaikunsopon and Phumkhachorn (2010) studied the 12 essential oils for antimicrobial activities against several strains of *Campylobacter jejuni*, a pathogen causing food-borne diseases worldwide. The authors showed that CEO exhibited the strongest antimicrobial activity against all the tested strains. The antimicrobial potency of coriander oil against *C. jejuni* on beef and chicken meat at 4 and 32 °C was also studied. It was found that the oil reduced the bacterial cell load in a dose-dependent manner; however, the type of meat and temperature did not influence the antimicrobial activity of the essential oil. This study clearly indicates the potential of the CEO to serve as a natural antimicrobial compound against *C. jejuni* in food.

In another study, the essential oils extracted from coriander and other plants were evaluated for their antimicrobial activity against 11 different bacterial and three fungal strains belonging to species reported to be involved in food poisoning and food decay. These include *S. aureus, E. coli, Salmonella enterica, L. monocytogenes, Bacillus cereus, C. albicans*, and *Aspergillus niger*. Coriander essential oil showed the best antibacterial activity in all, while thyme and spearmint oils better inhibited the fungal species (Lixandru et al., 2010). The antifungal activity of essential oils of some species of family *Apiaceae* was tested against the fungus *Aspergillus flavus*. Coriander oil was the most effective oil against fungal growth and aflatoxin production at all concentrations studied (Abou El-Soud et al., 2012). The authors recommended that an amount of 1000 ppm as a food additive is protecting the spices from bio-deteriorating fungi as well as from aflatoxin contamination.

Coriander oil expressed the highest antibacterial action against *E. coli* (10.73 ± 0.21) which was higher than gentamicin (9.47 ± 0.45). *E. coli* was followed by *Salmonella* (9.53 ± 0.40) which was slightly less than ampicillin (10.57 ± 0.21). Least activity of coriander was expressed against *Klebsiella* (7.20 ± 0.17) but that was quite near to action of ampicillin (8.43 ± 0.25). Coriander (*Coriandrum sativum L.*) essential oil showed apparently significant antibacterial activity against gram-negative microorganisms (*Escherichia coli, Klebsiella pneumoniae, Pseudomonas* and *Salmonella typhi*) and also gram-positive bacteria (*Staphylococcus aureus*). Further extensive studies are required to isolate and standardize the active phytoconstituent(s) responsible for antibacterial action and develop coriander essential oil as a clinically proven antibacterial agent (Sambasivaraju and Fazeel, 2016).

Rajeshwari and Andallu (2011) reported that coriander seeds contain petroselinic acid, linoleic acid, oleic acid and palmitic acid. Major components of essential oil are linalool, a-pinene, camphor and geraniol. They have demonstrated that these contents of coriander (also called cilantro) have some anti-bacterial action against *Salmonella* which is a frequent and at times the lethal cause of food poisoning and the activity of these compounds is comparable with gentamicin.

Additionally, the coriander leaf essential oil exhibited pronounced activity against Gram-positive and Gram-negative bacteria and their activity is quite comparable with the standard antibiotics such as tobramycin, gentamicin sulfate, ofloxacin and ciprofloxacin screened under similar conditions (Joji Reddy et al., 2012).

Therefore, the remarkable antibacterial activity exhibited by coriander essential oil could be attributed to the synergic effect of the active compounds present in it (Joji Reddy et al., 2012).

More recently, the synergy between essential oil components and antibiotics has been reviewed to highlight the possibilities of essential oils as the potential resistance modifying agent (Langeveld et al., 2014; Yap et al., 2014). Hence, interactive functions of the various essential oil components in comparison to the action of one or two main components of the oil seem unresolved. Furthermore, the whole essential oil exerts greater antibacterial activity compared to the major components alone (Burt, 2004). Thus, it is more reasonable to study the whole essential oil rather than some of its constituents as whether the concept of synergism truly exists between the components in essential oils (Bassolé and Juliani, 2012).

In this context, the synergistic antibacterial effect between coriander essential oil and conventional antibiotics against *Acinetobacter baumannii* was performed by checkerboard assays. Thus, the association of coriander essential oil with chloramphenicol, ciprofloxacin, gentamicin and tetracycline against *A. baumannii* showed *in vitro* effectiveness, which is an indicator of a possible synergistic interaction against two reference strains of *A. baumannii* (FIC index from 0.047 to 0.375). Nevertheless, an additive interaction was observed when tested the involvement between coriander essential oil and piperacillin or cefoperazone, the isobolograms. Consequently, the *in vitro* interaction could improve the antimicrobial effectiveness of ciprofloxacin, gentamicin and tetracycline and may contribute to resensitize *A. baumannii* to the action of chloramphenicol (Duarte et al., 2012).

Another research also suggests that the volatile oils found in the leaves of *C. sativum* plant may have antimicrobial properties against food borne pathogens such as *Salmonella* species (Isao et al., 2004). An aqueous decoction of coriander was found to have no bactericidal activity against *Helicobacter pylori* (O'Mahony et al., 2005). In contrast, some workers have found that *C. sativum* has strong antibacterial activity against both Gram-positive and Gram-negative (Al-Jedah et al., 2000).

ANTIBIOFILM ACTIVITY

The main objective of the Duarte et al. (2013) work was to determine the effect of coriander essential oil on *Acinetobacter baumannii* in different growth phases, as well as its ability to inhibit the formation or eradication of biofilms. The minimum inhibitory concentration (MIC) and minimum bactericidal concentration (MBC) of coriander oil using a microdilution broth susceptibility assay was determined. The effects of different concentrations of coriander oil (ranging from 0.125 to 4 × MIC) on biofilm formation and on established biofilm were tested using 96-well microtiter plates. Crystal violet assay was used as an indicator of total biofilm biomass and the biofilm viability was assessed with a XTT staining method. It was found that coriander oil presented significant antibacterial activity against all tested strains of *A. baumannii*, with MIC values between 1 and 4 μl/ml. The MBC values were the same as the MIC, being an indicator of the bactericidal activity of this essential oil. In what concerns the effect of this essential oil on biofilm formation inhibition was observed of at least 85% of biomass formation by all *A. baumannii* strains using 2 × MIC of coriander oil, in addition to a decrease in the metabolic activity of the cells. After exposure to coriander oil, a decrease in 24 h and 48 h-old biofilm biomass and the metabolism was seen for all tested concentrations, even with sub-inhibitory concentrations. Coriander essential oil proved to have a significant antibacterial and anti-biofilm activity and should be considered in the development of future disinfectants to control *A. baumannii* dissemination.

In the study of Freires et al. (2014), was aimed to investigate the antifungal activity and mode of action of the EO from *Coriandrum sativum* L. (coriander) leaves on *Candida* spp. In addition, was detected the molecular targets affected in whole-genome expression in human cells. The EO phytochemical profile indicates monoterpenes and sesquiterpenes as major components, which are likely to negatively impact the viability of yeast cells. There seems to be a synergistic activity of the EO chemical compounds as their isolation into fractions led to a decreased antimicrobial effect. *C. sativum* EO may bind to membrane ergosterol, increasing ionic permeability and causing membrane damage leading to cell death, but it does not act on cell wall biosynthesis-related pathways. This mode of action is illustrated by photomicrographs showing disruption in biofilm integrity caused by the EO at varied concentrations. The EO also inhibited *Candida* biofilm adherence to a polystyrene substrate at low concentrations and decreased the proteolytic activity of *Candida albicans* at minimum inhibitory concentration. Finally, the EO and its selected active fraction had low cytotoxicity on human cells, with putative mechanisms affecting gene expression in pathways involving chemokines and MAP-kinase (proliferation/apoptosis), as well as adhesion proteins. These findings highlight the potential antifungal activity of the EO from *C. sativum* leaves and suggest avenues for future translational toxicological research.

Minimum inhibitory concentration (MIC) of the *Coriandrum sativum* essential oils (EOs), dichloromethane extracts (DEs), and hexane extract (HE) were tested against *Candida* spp. planktonic cells. In general, the oils and extracts presented a good action against planktonic *Candida* spp. and clinical isolates; however, the oil and the hexane extract from *C. sativum* were capable of inhibiting all strains at MIC values between 0.015–0.500 mg/ml and 0.007–1.000 mg/ml. The activity of the fractions of *C. sativum* essential oil was also tested against the planktonic cells of *Candida* spp., presenting in general strong activity. *C. sativum* EO against *C. albicans* biofilm shows the kinetic biofilm development for *C. albicans* CBS 562 and clinical isolate 13A5. The action of nystatin, fluconazole, and *C. sativum* essential oil upon biofilm growth was presented. The crude essential oil from *C. sativum* and drugs was applied at regular intervals during biofilm cultivation. The results indicate a clear effect of the oil on biofilm formation, characterized due to an increase in the lag phase and a decrease in the biofilm growth. Similar effects were observed for nystatin, though inferior to fluconazole (Furletti et al., 2011).

In the study of Bazargani and Rohloff (2016) antiadhesion tests were carried out by crystal violet assay in order to evaluate essential oils and plant solvent extracts inhibition potential against cell attachments at MIC value concentration. Their results indicated that essential oil of coriander could inhibit bacteria cell attachment of *S. aureus* completely (100% inhibition activity), while the other extracts and essential oils generally displayed percentage inhibition in a range of 23-96%. *E. coli* was more resistant than *S. aureus* as observed and proved by lower percentage inhibition values. Inhibition of biofilm formation was conducted only on essential oils and solvent extracts, which showed at least 50% reduction (at MIC value concentration) in cell attachment on both tested bacteria by using crystal violet assay. Results of Bazargani and Rohloff (2016) showed different effects on the growth and development of a preformed biofilm. The essential oil of coriander induced inhibition of biofilm formation against *S. aureus* by up to 91% and 88.5%, respectively. While some solvent extracts increased biofilm growth and development of *S. aureus*, no inhibition was recorded for coriander. Results indicated that essential oil of coriander could inhibit biofilm formation of *E. coli* completely, displaying 100% inhibition activity. In comparison, the DCM extract of coriander did not prevent biofilm formation of *E. coli*. However, results indicated strong biofilm inhibition by coriander essential oil against *S. aureus* and *E. coli* when used at MIC value concentrations 0.8 and 1.6 ml/ml, respectively. The metabolic (respiratory) activity of cells in biofilm after 24 h was evaluated by using XTT reduction assay. The result indicated that most solvent extracts and essential oils reduced metabolic activity of cells in the biofilm of *S. aureus* and *E. coli*, showing an inhibition percentage range of 38.3-72.6% and 57.4-86%, respectively. In contrast, DCM and methanol extracts of coriander did not inhibit metabolic activity of biofilm cells of *S. aureus* at all (0%). When comparing all extracts and oils, essential oil of coriander was the most effective in inhibiting formation and growth of *S. aureus* biofilm by 72.6% and 71.5%. Their data also provided evidence that coriander oil had the highest inhibitory potential with 86% and 71.5% reduction in metabolic activity of *E. coli* and *S. aureus*, respectively, as it affected the oxidative activity of both tested bacteria. In summary, results from antiadhesion testing, crystal violet and XTT assays indicated that essential oils of coriander and methanol extract were effective in reducing biofilm biomass, and impaired metabolic activity of cells adherent in biofilm formed by *E. coli* and *S. aureus* (Bazargani and Rohloff, 2016).

ANTIMICROBIAL ACTIVITY OF CORIANDER SUBSTANCES

More recently, a novel antimicrobial peptide namely "Plantaricin CS" with a wide antibacterial activity was isolated from coriander leaf extract and the greatest antimicrobial effect of it was shown on *S. aureus* strain

(MIC = 1.3 mg/mL). Also, the new peptide showed effective germicidal effects on *K. pneumonia* (MIC = 2.65 mg/mL). However, *P. aeruginosa* was the most resistant strain (MIC = 3.2 mg/mL). Thus, Plantaricin CS has a less antimicrobial effect on Gram-negative bacteria which could be probably due to the presence of cell wall polysaccharides, preventing active compounds from reaching to the cytoplasmic membrane of these bacteria (Zare-Shehneh et al. 2014).

Aliphatic (2E)-alkenals and alkanals characterized from the fresh leaves of the coriander *Coriandrum sativum* L. (*Umbelliferae*) were found to possess bactericidal activity against *Salmonella choleraesuis* ssp. *choleraesuis* ATCC 35640. (2E)-Dodecenal (C_{12}) was the most effective against this food-borne bacterium with the minimum bactericidal concentration (MBC) of 6.25 µg/mL (34 µM), followed by (2E)-undecenal (C_{11}) with an MBC of 12.5 µg/mL (74 µM). The time-kill curve study showed that these α, β-unsaturated aldehydes are bactericidal against *S. choleraesuis* at any growth stage and that their bactericidal action comes in part from the ability to act as nonionic surfactants (Kubo et al. 2004).

Delaquis et al. (2002) examined the antibacterial activity of crude oil and the distilled fractions coriander (seeds of *C. sativum* L.) and cilantro (leaves of immature *C. sativum*) against some Gram-positive and Gram-negative food spoilage bacteria (*Salmonella typhimurium*, *Listeria monocytogenes*, *Staphylococcus aureus* (*S. aureus*), *Serratia grimesii*, *Enterobacter agglomerans*, *Yersinia enterocolitica*, *Bacillus cereus*). The inhibitory effect of *C. sativum* on potential spoilage bacteria, such as *Klebsiella pneumoniae* (*K. pneumoniae*), *Bacillus megaterium*, *Pseudomonas aeruginosa* (*P. aeruginosa*), *S. aureus*, *Escherichia coli* (*E. coli*), *Escherichia cloaca*, *Enterococcus faecalis*, has been reported (Keskin and Toroglu, 2011).

The *Coriander sativum* fruit EO (15 µl/disc) exhibited an antibacterial effect against *E. coli*, *P. aeruginosa* and *Salmonella typhi* (*S. typhi*) showing zone diameter of inhibition 25, 10 and 18 mm, respectively (Teshale et al., 2013); at such levels, the modes of action of oils of *C. sativum* were demonstrated to be bactericidal against *S. typhi* and bacteriostatic action against *E. coli*. Linalool, the major component of the oil, which was reported to have an antibacterial effect against many bacterial strains (Ates and Erdorul 2003), could be responsible for antibacterial activity. As has been reported by Lalitha et al.

(2011), the concentration dependant antibacterial activity of the CEO against potential food poisoning bacteria causing serious infection to humans.

Plantaricin *C. sativum*, an antimicrobial peptide containing 26 amino acids, isolated from coriander leaf extract exerted antimicrobial activities against Gram-negative bacteria showing minimum inhibitory concentration (MIC) values 71.55 and 86.4 µg/mL, respectively for *K. pneumoniae* and *P. aeruginosa* as well as Gram-positive (MIC 35.2 µg/mL for *S. aureus*) bacteria (Zare-Shehneh et al., 2014).

Nanasombat and Lohasupthawee (2005) demonstrated the inhibitory effect of *C. sativum* EO against 25 bacterial strains (20 serotypes of *Salmonella* and 5 species of other enterobacteria: *Citrobacter freundii, Enterobacter aerogenes, E. coli, K. pneumoniae*, and *Serratia marcescens* showing MIC of 4.2 µL/mL to most bacterial strains; *Salmonella enterica* serotype *risen* was resistant strains to *C. sativum* EO (>62.5 µL/mL). Innocent *et al. (2011)* evaluated the immunostimulant potential of *C. sativum* in fish *Catla catla*, post challenged with *Aeromonas hydrophila*, and thus, it has been found to be a good choice as a diet supplement to induce disease resistance in fishes.

For *Candida* spp., MICs of CLEO ranged 15.6–31.2 µg/ml, and minimum fungicidal concentrations (MFCs) 31.2–62.5 µg/ml, while the active fraction had higher MIC and MFC values, ranging from 31.2 µg/ml to 250 µg/ml and 125 µg/ml to 1000 µg/ml, respectively, indicating a synergistic activity of the EO components (mono- and sesquiterpene hydrocarbons) as their isolation into fractions led to a decreased antimicrobial effect, and hence *C. sativum* EO can be used as a potential candidate in the treatment of oral diseases, such as denture-related candidiasis (Freires et al., 2014). The CEO showed excellent antifungal activity against seed borne pathogens of paddy *Pyricularia oryzae, Bipolaris oryzae, Alternaria alternata, Tricoconis padwickii, Drechslera tetramera, Drechslera halodes, Curvularia lunata, Fusarium moniliforme* and *Fusarium oxysprorum* (Lalitha *et al. 2011).* The concentration-dependent killing activity of coriander oil for *P. oryzae, T. padwickii, B. oryzae, A. alternata, D. halodes* and *F. oxysprorum*. Zare-Shehneh *et al. (2014)* showed that the plantaricin *C. sativum*, from coriander leaf extract, had fungicidal activity against *Penicillium lilacinum* and *Asperjilus niger* with MICs 67.8 and 62.1 µg/ml, respectively.

USES OF APPLICATIONS

The coriander plant is a highly valued medicinal plant. All parts of the plant are in use as traditional remedies for the treatment of different disorders in the folk medicine systems of different civilizations. It is a potential source of lipids (rich in petroselinic acid) and essential oil (high in linalool) isolated from the seeds and the aerial parts. Due to the presence of several bioactive ingredients, wide arrays of pharmacological activities have been ascribed. A review by Sahib et al. (2013) concluded the antimicrobial, antioxidant, antidiabetic, anxiolytic, antiepileptic, anti-depressant, antimutagenic, anti-inflammatory, antidyslipidemic, antihyper-tensive, neuroprotective, and diuretic properties of coriander may be ascribed to the bioactive components present in essential oil. Similarly, Ramadan et al. (2008) demonstrated that the CEO has hypocholesterolemic properties in rats fed a cholesterol-rich diet.

CONCLUSION

In conclusion, several industries (pharmacy, medicine, and food technology) are now looking for sources of new, natural, antimicrobial and safety bioactive compounds. The extracts and essential oils from *Coriandrum sativum* L. showed in many research and clinical studies antioxidant, antimicrobial, anti-inflammatory, anticancer and cytotoxic efficacy. These extracts and essential oils can also retard lipid oxidation in food matrices, so they can be used in future in food technology as potential substitutes of synthetic antioxidants in food preservation. The essential oil also contributes to the storage stability of the food products. Coriander essential oil possesses strong potential against food-spoiling microorganisms. The dietary uses of the CEO are therefore helpful in maintaining good health.

REFERENCES

Abou El-Soud, N.H., Deabes, M.M., Abou El-Kassem, L.T., Khalil, M.Y., 2012. Antifungal activity of family apiaceae essential oils. *Journal of Applied Science and Research* 8: 4964-4973.

Alboofetileh, M. Rezaei, M. Hosseini, H. Abdollahi, M. 2014. Antimicrobial activity of alginate/clay nano-composite films enriched with essential oils against three common foodborne pathogens. *Food Control* 36: 1-7.

Al-Jedah, J.H. Ali, M.Z. Robinson, R.K. 2000. The inhibitory action of spices against pathogens that might be capable of growth in a fish sauce (Mehiawah) from the Middle East. *International Journal of Food Microbiology* 57: 129-133.

Al-Snafi, A.E. 2016. A review on chemical constituents and pharmacological activities of *Coriandrum sativum*. *IOSR Journal of Pharmacy* 6: 17-42.

Ates, D.A. Erdogrul, O.T. 2003. Antimicrobial activities of various medicinal and commercial plant extracts. *Turkish Journal of Biology* 27: 157-162.

Bassolé, I.H.N. Juliani, H.R. 2012. Essential oils in combination and their antimicrobial properties. *Molecules* 17(4): 3989-4006.

Bazargani, M.M. Rohloff, J. 2016. Antibiofilm activity of essential oils and plant extracts against *Staphylococcus aureus* and *Escherichia coli* biofilms. *Food Control* 61:156-164.

Bhat, S. Kaushal, P. Kaur, M. Sharma, HK. 2014. Coriander (*Coriandrum sativum* L.): Processing nutritional and functional aspects. *African Journal of Plant Science* 8: 25-33.

Burt, S. 2004. Essential oils: their antibacterial properties and potential applications in foods—a review. *International Journal of Food Microbiology* 94: 223-253.

Cantore, P.L. Iacobellis, N.S. De Marco, A. Capasso, F. Senatore, F. 2004. Antibacterial activity of *Coriandrum sativum* L. and *Foeniculum vulgare* Miller Var. vulgare (Miller) essential oils. *Journal of Agricultural and Food Chemistry* 52: 7862-7866.

Cao, X.Z. You, J.M. Shen-Xin Li, S.X. Zhang, Y.L. 2012. Antimicrobial Activity of the Extracts from *Coriandrum sativum*. *International Journal of Food Nutrition and Safety* 1(2): 54-59.

Casetti, F. Bartelke, S. Biehler, K. Augustin, M. Schempp, C.M. Frank, U. 2012. Antimicrobial activity against bacteria with dermatological relevance and skin tolerance of the essential oil from *Coriandrum sativum* L. fruits. *Phytotheraphy Research* 26(3): 420-424.

Charles, D.J. 2013. *Antioxidant properties of spices, herbs and other sources*. USA: Springer, pp. 256-258.

Chizoba, E.F.G. 2015. A comprehensive review on coriander and its medicinal properties. *International Journal of Scientific Research and Reviews* 4: 28-50.

Darughe, F. Barzegar, M. Sahari, M.A. 2012. Antioxidant and antifungal activity of coriander (*Coriandrum sativum* L.) essential oil in cake. *International Food Research Journal* 19: 1253-1260.

Deb Roy, S. Thakur, S. Negi, A. Kumari, M. Sutar, N. Kumar Jana, G. 2010. *In vitro* antibiotic activity of volatile oils of *Carum carvi* & *Coriandrum sativum*. *International Journal of Chemistry and Analytical Sciences* 1(7): 149-150.

Deepa, B. Anuradha, CV. 2011. Antioxidant potential of *Coriandrum sativum* L. seed extract. *Indian Journal of Experimental Biology* 29: 53-56.

Delaquis, P.J. Stanich, K. Girard, B. Mazza, G. 2002. Antimicrobial activity of individual and mixed fractions of dill, cilantro, coriander and eucalyptus essential oils. *International Journal of Food Microbiology* 74: 101-109.

Dua, A. Garg, G. Kumar, D. Mahajan, R. 2014. Polyphenolic composition and antimicrobial potential of methanolic coriander (*Coriandrum sativum*) seed extract. *International Journal of Pharmacology Science and Research* 5(6): 2302-08.

Duarte, A. Ferreira, S. Silva, F. Domingues, F.C. 2012. Synergistic activity of coriander oil and conventional antibiotics against *Acinetobacter baumannii*. *Phytomedicine* 19(3-4): 236-238.

Duarte, A. Ferreira, S. Oliveira, R. Domingues, F. 2013. Effect of Coriander Oil (Coriandrum sativum) on Planktonic and Biofilm Cells of *Acinetobacter baumannii*. *Natural product communications* 8(5):673-678.

Duman, A.D. Telci, I. Dayisoylu, K.S. Digrak, M. Demirtas, I. Alma, M.H. 2010. Evaluation of bioactivity of linalool-rich essential oils from *Ocimum basilucum* and *Coriandrum sativum* varieties. *Natural Products Communication* 5(6): 969-974.

Fayyad, A.G. Yaakobi, A.A. 2017. Evaluation of biological activities of seeds of *Coriandrum sativum*. *International Journal of Scientific and Engineering Research* 8: 1058-1064.

Freires, I.D.A. Murata, R.M. Furletti, V.F. Sartoratto, A. de Alencar, S.M.D. Figueira, G.M. 2014. *Coriandrum sativum* L. (Coriander) essential oil: antifungal activity and mode of action on *Candida* spp., and molecular targets affected in human whole-genome expression. *PLoS One* 9: e99086.

Furletti, V.F. Teixeira, I.P. Obando-Pereda, G. Mardegan, R.C. Sartoratto, A. Figueira, G.M. Duarte, R.M.T. Rehder, V.L.G. Duarte, M.C.T. Höfling, J.F. 2011. "Action of Coriandrum sativum L. Essential Oil upon Oral *Candida albicans* Biofilm Formation." *Evidence-Based Complementary and Alternative Medicine*, 2011: Article ID 985832.

Gupta, H. 2019. *An overview on family –Apiaceae (Umbelliferae) botany* Available from: http://www.biologydiscussion.com/plants/floweringplants/an-overview-on-family-apiaceae-umbelliferae-botany/19892.

Hwang, E. Lee, D.G. Park, S.H. Oh, M.S. Kim, S.Y. 2014. Coriander leaf extract exerts antioxidant activity and protect against UVB-induced photoaging of skin by regulation of procollagen type I and MMP-1 expression. *Journal of Medicinal Foods* 17: 1-10.

Ildiz, N. Baldemir, A. Ince, U. Ilgun, S. Konca, Y. 2018. Synergistic effect of *Coriandrum sativum* L. extracts with cefoxitin against methicillin resistant *Staphylococcus aureus*, extended-spectrum beta-lactamase producing *Escherichia coli* and *Klebsiella pneumonia*. *Medicine Science* 7(4):777-780.

Innocent, B.X. Fathima, Dhanalakshmi, M.S.A. 2011. Studies on the immouostimulant activity of *Coriandrum sativum* and resistance to *Aeromonas hydrophila* in *Catla catla*. *Journal of Applied Pharmacology Sciences* 1: 132-135.

Isao, K. Ken-Ichi, K.Aya, F. Ken-Ichi, N. Tetsuya, A. 2004. Antibacterial activity of coriander volatile compounds against Salmonella choleraesuits. *Journal of Agriculture and Food Chemistry* 52(11): 3329-3332.

Joji Reddy, L. Devi Jalli, R. Jose, B. Gopu, S. 2012. Evaluation of antibacterial and DPPH radical scavenging activities of the leaf extracts and leaf essential oil of *Coriandrum sativum* Linn. *World Journal of Pharmacology Research* 1(3): 705-716.

Keskin, D. Toroglu, S. 2011. Studies on antimicrobial activities of solvent extracts of different spices. *Journal of Environmental Biology* 32: 251-256.

Koppula, S. Choi, D.K. 2012. Antistress and anti-amnesic effects of *Coriandrum sativum* L. (*Umbelliferae*) extract – an experimental study in rats. *Tropical Journal of Pharmaceutical Research* 11: 36-42.

Kubo, I. Fujita, K.I. Kubo, A. Nihei, K.I. Ogura, T. 2004. Antibacterial activity of coriander volatile compounds against *Salmonella choleraesuis*. *Journal of Agricultural and Food Chemistry* 52: 3329-3332.

Lalitha, V. Kiran, B. Raveesha, K.A. 2011. Antifungal and antibacterial potentiality of six essential oils extracted from plant source. *International Journal of Engineering Science and Technology* 3: 3029-3038.

Langeveld, W.T. Veldhuizen, E.J. Burt, S.A. 2014. Synergy between essential oil components and antibiotics: a review. *Critical Revue of Microbiology* 40(1): 76-94.

Laribi, B. Kouki, K. M'hamdi, M. Bettaieb, T. 2015. Coriander (*Coriandrum sativum* L.) and its bioactive constituents. *Fitoterapia* 103: 9-26.

Lixandru, B.E. Drăcea, N.O. Dragomirescu, C.C. Drăgulescu, E.C. Coldea, I.L. Anton, L. 2010. Antimicrobial activity of plant essential oils against bacterial and fungal species involved in food poisoning and/or food decay. *Roumanian Archives of Microbiology and Immunology* 69(4): 24-230.

Mahendra, P. Bisht, S. 2011. Anti-anxiety activity of *Coriandrum sativum* assessed using different experimental anxiety models. *Indian Journal of Pharmacology* 43: 574-577.

Marangoni, C. Moura, N.F. 2011. Antioxidant activity of essential oil from *Coriandrum sativum* L. in Italian salami. *Food Science and Technology* 31: 124-128.

Matasyoh, J.C. Maiyo, Z.C. Ngure, R.M. Chepkorir, R. 2009. Chemical composition and antimicrobial activity of the essential oil of *Coriandrum sativum. Food Chemistry* 113: 526–529.

Misharina, T.A. Polshkov, A.N. 2005. Antioxidant properties of essential oils: autoxidation of essential oils from Laurel and Fennel and of their mixtures with essential oil from coriander. *Applied Biochemical Microbiology* 41: 610-618.

Msaada, K. Jemia, M.B. Salem, N. Bachrouch, O. Sriti, J. Tammar, S. Bettaieb, I. Jabri, I. Kefi, S. Limam, F. Marzouk, B. 2017. Antioxidant activity of methanolic extracts from three coriander *Coriandrum sativum* L. fruit varieties. *Arabian Journal of Chemistry* 10: 3176-3183.

Nadeem, M. Anjum, F.M. Khan, M. Tahseen, S. 2013. Nutritional and medicinal aspects of coriander (Coriandrum sativum L.): A review. *British Food Journal* 115: 743-755.

Nanasombat, S. Lohasupthawee, P. 2005. Antibacterial activity of crude ethanolic extracts and essential oils of spices against salmonellae and other enterobacteria. *KMITL Sciences Technology Journal* 5: 527-538.

Nithya, T.G. Sumalatha, D. 2014. Evaluation of *in vitro* antioxidant and anticancer activity of *Coriandrum sativum* against human colon cancer HT-29 cell lines. *International Journal of Pharmacy and Pharmaceutical Sciences* 62: 421-424.

O'Mahony, R. Al-Khtheeri, H. Weerasekera, D. Fernando, N. Vaira, D. Holton, J. Basset, C. 2005. Bactericidal and anti-adhesive properties of culinary and medicinal plants against Helicobacter pylori. *World Journal of Gastroenterology* 11(47): 7499-7507.

Önder, A. 2018. Coriander and its phytoconstituents for the beneficial effects. In El-Shemy, H. 2018. *Potential of Essential Oils*. IntechOpen: London, p. 166.

Pandey, A. Bigoniya, P. Raj, V. Patel, K.K. 2011. Pharmacological screening of *Coriandrum sativum* Linn. for hepaprotective activity. *Journal of Pharmacological Bioallied Science* 3: 435-441.

Pawar Vinita, A. Bhagat Trupti, B. Toshniwal Mitesh, R. Mokashi Nitin, D. Khandelwal, K.R. 2013. Formulation and evaluation of dental gel containing essential oil of coriander against oral pathogens. *International Research Journal of Pharmacology* 4 (10): 48-54.

Rajeshwari, U. Andallu, B. 2011. Medicinal benefits of coriander. *Spatula DD* 1(1):51-8.

Ramadan, M.F. Amer, M.A.A. Awad, A.E. 2008. Coriander (*Coriandrum sativum* L.) seed oil improves plasma lipid profile in rats fed a diet containing cholesterol. *European Food Research and Technology* 227: 1173-1182.

Ramadan, M.F. Wahdan, K.M.M. 2012. Blending of corn oil with black cumin (*Nigella sativa*) and coriander (*Coriandrum sativum*) seed oils: impact on functionality, stability and radical scavenging activity. *Food Chemistry* 132: 873-879.

Rattanachaikunsopon, R. Phumkhachorn, P. 2010. Potential of coriander (*Coriandrum sativum*) oil as a natural antimicrobial compound in controlling *Campylobacter jejuni* in raw meat. *Bioscience, Biotechnology and Biochemistry* 74: 31-35.

Saeed, M. Amen, A. Fahmi, A. Garawan, IE. Sayed, S. 2014. The possible protective effect of *Coriandrum sativum* seeds methabolic extract on hepato.renal toxicity induced by sodium arsenic in albino rats. *Journal of Applied Pharmaceutical Science* 4: 44-51.

Sahib, N.G. Anwar, F. Gilani, A.H. Hamid, A.A. Saari, N. Alkharfy, K.M. 2013. Coriander (*Coriandrum sativum* L.): a potential source of high-value components for functional foods and nutraceuticals- a review. *Phototherapy Research* 27: 1439-1456.

Sambasivaraju, D. Fazeel Z.A. 2016. Evaluation of antibacterial activity of *Coriandrum sativum (L.)* against gram – positive and gram – negative bacteria. *International Journal of Basic and Clinical Pharmacology* 5(6): 2653-2656.

Sharma, V. Kansal, L. Sharma A. 2010. Prophylactic efficacy of *Coriandrum sativum* (coriander) on testis of lead-exposed mice. *Biology Trace Element Research* 136: 337-354.

Siddiqi, A. Parveen, A. Dhyani, N. Hussain, ME. Fahim, M. 2017. Effects of *Coriandrum sativum* extract and simvastatin in isoproterenol induced heart failure in rats. *Serbian Journal of Experimental and Clinical Research* 19: 1198-129.

Silva, F. Ferreira, S. Queiroz, J.A. Domingues, F.C. 2011. Coriander (*Coriandrum sativum* L.) essential oil: its antibacterial activity and mode of action evaluated by flow cytometry. *Journal of Medical Microbiology* 60: 1479-1486.

Talaei, R. Kheirollah, A. Rezaei, HB. Mansouri, E. Mohammadzadeh, G. 2017. Protective effects of hydro-alcoholic extract of *Coriandrum sativum* in rats with experimental iron-overload condition. *Jundishapur Journal of Natural Pharmaceutical Products* 12: 1-9.

Tang Esther, L.H. Rajarajeswaran, J. Fung, S.Y. Kanthimathi, M.S. 2013. Antioxidant activity of *Coriandrum sativum* and protection against DNA damage and cancer cell migration. *Complementary and Alternative Medicine* 2: 347.

Teshale, C. Hussien, J. Jemal, A. 2013. Antimicrobial activity of the extracts of selected Ethiopian aromatic medicinal plants. *Spatula DD* 3: 175-180.

Thompson, A. Meah, D. Ahmed, N. Conniff-Jenkins, R. Chileshe, E. Phillips, C.O. 2013. Comparison of the antibacterial activity of essential oils and extracts of medicinal and culinary herbs to investigate potential new treatments for irritable bowel syndrome. *BMC Complementary and Alternative Medicine* 13: 338.

Usta, J. Kreydiyyeh, S. Knio, K. Barnabe, P. Bou-Moughlabay, Y. Dagher, S. 2009. Linalool decrease HepG2 viability by inhibiting mitochondrial complexes I and II, increasing reactive oxygen species and decreasing ATP and GSH levels. *Chemico-Biological Interaction* 180: 39-46.

Vejdani, R. Shalmani, H.R.M. Mir-Fattahi, M. Sajed-Nia, F. Abdollahi, M. Zali, M.R. 2006. The efficacy of an herbal medicine, Carmint, on the relief of abdominal pain and bloating in patients with irritable bowel syndrome: a pilot study. *Digestive Disease and Sciences* 51(8): 1501-1507.

Verma, A. Pandeya, S.N. Sanjay, K.Y. Styawan, S. 2011. A Review on *Coriandrum sativum* (Linn.). An Ayurvedic medicinal herb of happiness. *Journal of Advanced Pharmaceutical Healthcare Research* 1: 28-48.

Wu, T.T. Tsai, C.W. Yao, H.T. Lii, C.K. Chen, H.W. Wu, Y.L. Chen, P.Y. Liu, K.L. 2010. Suppressive effects of extracts from the aerial part of *Coriandrum sativum* L. on LPS-induced inflammatory responses in murine RAW 264.7 macrophages. *Journal of the Science of Food and Agriculture* 90: 1846-1854.

Yadav, N.P. Luqman, S. Meher, J.G. Sahu, A.K. 2012. Effect of different pharmaceutical vehicles on antimicrobial action of essential oils. *Planta Medica* 78(11): PF44.

Yap, P.S.X. Yiap, B.C. Ping, H.C. Lim, S.H.E. 2014. Essential oils, a new horizon in combating bacterial antibiotic resistance. *Open Microbiology Journal* 8: 6-14.

Yildiz, H. 2016. Chemical composition, antimicrobial and antioxidant activities of essential oil and ethanol extract of *Coriandrum sativum* L. leaves from Turkey. *International Journal of Food Properties* 19: 1593-1603.

Zare-Shehneh, Z. Askarfarashah, M. Ebrahimi, L. Moradi Kor, N. Zare-Zardini, H. Soltaninejad, H. 2014. Biological activities of a new antimicrobial peptide from *Coriandrum sativum*. *International Journal of Biosciences* 4(6): 89-99.

Zare-Zardini, H. Tolueinia, B. Momeni, Z. Hasani, Z. Hasani, M. 2012. Analysis of antibacterial and antifungal activity of crude extracts from seeds of *Coriandrum sativum*. *Gomal Journal of Medicinal Sciences* 10(2): 167-171.

In: Coriander
Editor: Deepak Kumar Semwal

ISBN: 978-1-53616-483-1
© 2019 Nova Science Publishers, Inc.

Chapter 4

CORIANDRUM SATIVUM:
A PLANT OF HEALTH BENEFITS AND BIOTECHNOLOGICAL APPLICATIONS FOR IMPROVEMENT

Muzamil Ali, A. Mujib[*], *Nadia Zafar and Basit Gulzar*
Cellular Differentiation and Molecular Genetics Section,
Department of Botany, Jamia Hamdard, New Delhi, India

ABSTRACT

Coriandrum sativum L. is an annual herb belonging to the family Apiaceae. The plant is used as a vegetable as it is a rich source of vitamin A, B_2 and C. The whole plant and the unripe fruits possess strong odour with a characteristic aroma. The fresh green herb and seeds are the primary coriander products, used as a spice and essential oil in the aroma and flavour industry. It has traditionally been used against antidiabetic, antioxidant, anti-inflammatory, antibacterial, antifungal and antitumor activities. The flavour of coriander is determined by the composition of the oil, and linalool is one important component found more in ripened fruits

---

[*]Corresponding Author's Email: amujib3@yahoo.co.in

than that of other parts. Recently, the biotechnological techniques became more popular in enriching medicinally active compounds and in preserving endangered plants/cell lines by conserving germplasm. Plant tissue culture made the plants available all year round by preserving important genetic materials including coriander. The success of *in vitro* propagation (organogenesis and embryogenesis) and improvement of coriander is possible through various biotechnological ways like genetic transformation and somatic hybridization in which embryo precursor cells can play a very pivotal role. These *in vitro* raised embryos mimic zygotic embryos and thus can be useful as a model for studying the fundamentals of plant development, differentiation and cellular totipotency especially at the morphological, biochemical and molecular level. The process is also widely exploited to obtain genetically uniform, disease-free plants and in achieving better-adapted cultivars with high multiplication rate, resistant to biotic and abiotic stresses. The isolated protoplasts of two or more sources of different traits are an important experimental system in molecular breeding especially in identifying physiological, biochemical and genetical sexual incompatibility. Exploiting the unique properties, protoplasts could create new germplasm via somatic hybridization in coriander. In the present chapter, the medicinal importance/qualities of coriander have been summarized and some of the different biotechnological ways of improvement have also been discussed in coriander.

Keywords: germplasm, embryogenesis, biotechnology, *in vitro* tissue culture

INTRODUCTION

Natural vegetation has been an integral part of human life since its origin. In addition to food, we are dependent on natural vegetation for various medicinal purposes (Micke et al., 2009). Plants produce secondary metabolites which serve important source to the pharmaceutical industry for the manufacturing of drugs (Oksman-Caldentey and Inze, 2004). At the same time, the indiscriminate usage of medicinal plants in ethnobotany resulted in habitat degradation and loss of genetic diversity posing threat to their extinction (Edwards, 2004). The alternative method to the conservation of wild plant resources with simultaneous fulfilment of present demands of medicine can be fulfilled by adopting *in vitro* culture technique. Plant tissue culture made the availability of plant materials possible at any time by preserving the genetic material of important plant species including the coriander. It is an *in vitro* aseptic technique based on the principle of totipotency for the growth and multiplication of plant cells, tissues, organs and whole plants under conducive nutritional and environmental conditions to regenerate plants (Thorpe, 2007). The success of the process is largely dependent on three factors: explant choice, medium composition and the physical environment (El-Meskaoui, 2013). The composition of a nutrient medium, suitable to a certain plant species is the main task for the establishment of culture. The biotechnology has been developed to such an extent that any plant species can now be *in vitro* regenerated through one of the following methods: embryo-, anther-culture, callogenesis, somatic or asexual embryogenesis and organogenesis (Ahsan et al., 2014). The choice of method primarily depends on plant species and the method of producing plants at a realistic cost (Kumar et al., 2012). The technology is being widely used in plant propagation, plant improvement, disease elimination and production of secondary metabolites (Hussain et al., 2012a). Commercial production of plants through micropropagation has several advantages over the traditional methods of propagation (El-Dougdoug and El-Sham, 2011).

Plant biotechnology is considered to be a way of producing new secondary metabolites with improved biological activities that will have value to the pharmaceutical industry in providing primary healthcare facilities at a lower cost (Hussain et al., 2012b). Many plant species containing high-value compounds are difficult to cultivate and the chemical synthesis of plant-derived compounds is often not economically feasible because of their highly complex structures (Ji et al., 2009). The lower quantity of secondary metabolites under natural conditions has led researchers to develop production technology to enhance and harvest pharmaceutically active metabolites under *in vitro* (Hussain et al., 2012b). The successful production of secondary metabolites from cell cultures of various plant species under *in vitro* conditions resulted in safeguarding economically important endangered plant germplasm (Guo et al., 2007).

Medicinal plants play a key role in the development and advancement of modern studies by serving as a starting material for the development of novel drugs (Pramono, 2002). The 25% of drugs used in modern

Pharmacopoeia against diverse diseases like convulsion, anxiety, insomnia and loss of appetite (Rao et al., 2004) are derived from estimated 2500 species of medicinal plants (Emamghoreishi et al.,2005). Medicinal plants produce rich diversity of secondary metabolites in very small quantities under wild condition, while their consumption is increasing due to the majority of people relying on traditional medicine (Kala et al., 2006). This need for their enhanced quantities hence met by *in vitro* cell cultures while the advancement in biotechnology has further improved this potential as an essential source of secondary metabolites (Rao and Ravishankar, 2002). The *in vitro* culture of medicinal and aromatic plants can provide an alternative way of consistent isolation of medicinal compounds from plant materials (Smetanska, 2008). Different strategies have been employed to improve the production of these compounds (Szopa et al., 2012), these include various types of stress conditions that enhance the production of secondary metabolites (Khan et al., 2011). Stress conditions reduce plant growth but stimulate the synthetic rate of highly reduced compounds due to an oversupply of reduction equivalents, with optimization of growth conditions for biomass production and the synthesis and accumulation of metabolites (Hakkinen and Ritala, 2010). Consequently, using an elicitor could initiate or improve the biosynthesis of secondary metabolites under *in vitro* conditions (Smetanska, 2008).

APPLICATIONS OF CORIANDER

C. sativum is grown as an annual herb and belongs to family Apiaceae. There are two species in genus *Coriandrum*: *C. tordylium* and *C. sativum*, former is a wild species and latter is a cultivated one. The spice is a diploid cross-pollinated crop which is native to southern Europe and the western Mediterranean region and presently worldwide in cultivation (Innocent et al., 2011). *C. sativum*, on the basis of seed size, has been classified into two main groups: small-seeded varieties (microcarpum) are cultivated in temperate climates such as Russia, and eastern and central European countries while large-seeded varieties (vulgare or macrocarpum) are widely grown in tropical and subtropical climates (Small, 1997).

The oil obtained from coriander (coriander oil) is listed among 20 major essential oils in the world market (Lawrence, 1992). The commercial value of coriander oil principally depends on its aroma, chemical composition and physical properties (Smallfield et al., 2001). The essential oil predominated by oxidized monoterpenes like linalool and monoterpene hydrocarbons and is used for aroma and flavour industry (Bhuiyan et al., 2009). The volatile oil of fresh herb also contains aliphatic aldehydes (mainly C_{10}-C_{16}) as main components (Potter and Fagerson, 1990).

Fatty acids such as palmitic-, petroselenic-, linoleic-, and stearic acids are also present in coriander seeds (Ramadan and Morsel, 2003). The whole plant body and in particular the unripe fruits possess strong odour with characteristic aroma (Pathak et al., 2011). The dried coriander seeds contain an essential oil (0.03% to 2.6%) with linalool as the main component (Coskuner and Karababa, 2007; Eikani et al., 2007), phenolics, flavonoids (Helle et al., 2004) and isocoumarin compounds (Taniguchi et al., 1996). The plant is used as a vegetable as it is a rich source of vitamin A, B_2 and C (Prakash, 1990). Coriander has many medicinal activities including antidiabetic (Gray and Flatt, 1999; Naquvi, Ali and Ahamad, 2012) and cholesterol-lowering effect (Chithra and Leelamma, 1997; Lal et al., 2004). *C. sativum* is used as an antioxidant (Darughe et al., 2012), antiinflammatory (Neha et al., 2013), antimutagenic (Cortes et al., 2004), antibacterial (Matasyoh et al., 2009), antifungal (Silva et al., 2011) and in antitumor activity (Chithra and Leelama, 2000). *C. sativum* has also been recommended for dyspeptic complaints, convulsion, anxiety, insomnia and loss of appetite (Emamghoreishi et al., 2005).

The essential oil quality and quantity is determined by several factors including genetic (Small 1997) and environmental conditions (Fuente et al., 2003). The oil content varies in different coriander varieties and the level ranges from 0.10-1.8% (Small, 1997; Fuente et al., 2003). The odour and flavour of coriander are determined by the composition of the oil (Bandoni et al., 1998). It is well known that linalool content in the essential oil of ripened coriander fruits is higher than that of mature fruits (Diederichsen,

1996). In order to cultivate a high-quality product, it is important to determine the effect of different ecological conditions on cultivated crops (Tilman et al., 2002). Despite extensive studies on oil components in coriander, the literature on variations of oil composition of regenerated *C. sativum* varieties under *in vitro* culture conditions is lacking.

IN VITRO PROPAGATION AND PRESERVATION

Biotechnological methods are important in the multiplication and improvement of economically important plants in general and coriander in particular, under *in vitro* conditions. Plant cell/tissue culture technique is being widely used for large scale plant propagation, besides being an *in vitro* experimental model for studying key processes of plant development like cell division, differentiation and cellular morphogenesis (Zimmerman, 1993, Gulzar et al., 2019). Cellular totipotency is central in plant cell-culture based regeneration process wherein plasticity allows differentiated plant cells to alter their metabolism, growth and development to suit cultural environment (Ondrej et al., 2009). Under *in vitro* conditions, coriander explants have the ability to initiate cell division, undergo different developmental pathways in response to particular stimuli in regenerating whole plants. It depends upon the concept that plant cells upon stimuli express the total genetic potential of the parent plant thereby maintaining the genetic potential called 'totipotency'. We observed the involvement of 2, 4-D for callus induction and proliferation in coriander explants including hypocotyl, root and cotyledon. The concentrations and combinations of different PGRs can also significantly cause differences in morphogenetic responses of specific genotypes, even among explants within a plant (Delporte et al., 2014). This also reconfirms that different PGRs are required at successive stages of *in vitro* morphogenesis and differentiation process. The acquisition of embryogenic competence under *in vitro* culture conditions largely depends on the growth stage at which the cells become responsive to new signals in altering metabolism (Sharmin et al., 2014). Cellular competence is associated with dedifferentiation of somatic cells that allows them to respond to new developmental signals by some specific characteristics such as early activation of the division cycle, more alkali vascular pH, an altered auxin metabolism, and a non-functional chloroplast (Potters et al., 2007). During cultural conditions 2, 4-D increases the concentration of endogenous auxin levels in cultured cells to determine their fate to become embryogenic (Jimenez and Thomas, 2005). The callus initiation being the first stage of differentiation from the parental tissue is essential in generating variability to introduce new desirable traits and generating transgenic plants (Zheng et al., 2005). In *C. sativum* this induction phase requires the presence of 2, 4-D in the culture medium to start the embryogenic process (Figure 1). After induction of somatic embryogenesis, the persistent presence of 2, 4-D hinders embryo development and maturation; hence the transfer of embryogenic cultures onto 2, 4-D free medium is suggested to be necessary for further development. Somatic embryo differentiation and maturation was also high in NAA and BA added medium. Finally, at later stages of embryogenesis/ maturation, the embryos desiccate and accumulate reserve nutrients for germination. The early somatic embryos/embryogenic tissues were thus transferred to NAA and BA supplemented medium where development and maturation were observed. In the process of somatic embryogenesis somatic cells undergo morphological and metabolic changes; acquire embryogenic potency, which later develops into somatic embryos (Feher et al., 2003).

In several ways, somatic embryos mimic zygotic embryos and are a useful model for studying the fundamentals of plant development and cellular totipotency especially at morphological and biochemical level (Santos et al., 2005). The acquisition of embryogenic competence via dedifferentiation and gene expression reprogramming due to chromatin remodelling in somatic cells are prerequisites for somatic embryogenesis (Verdeil et al., 2007). The rate of regeneration is high in callus phase compared to direct embryogenesis method, which serves as an efficient mass propagation means of commercially important plant species and conservation of rare, threatened, and endangered germplasm (Rao, 2004). The regenerated plants are disease free, genetically uniform and better-adapted cultivars with high multiplication rate (Medrano et al., 2014). The final stage in the regeneration process involves germination of mature somatic embryos on MS supplemented with BA and GA_3 into plantlets. In order to promote root growth in coriander regenerates, the well-developed plantlets were transferred to half strength MS amended with IBA.

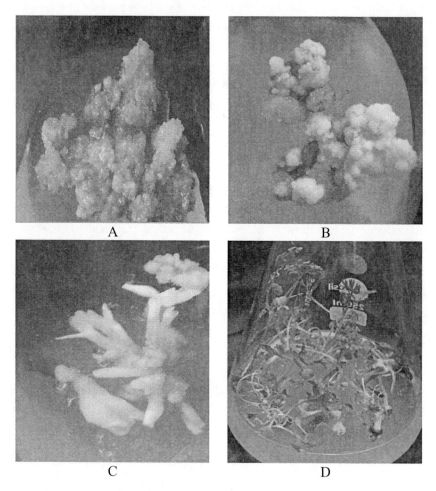

Figure 1. Callus induction, somatic embryogenesis and plant regeneration in coriander; (A) callus stage; (B, C) Different stages of embryos; (D) plant regeneration.

During SE, it was observed that the external factors are necessary for coordinating cell division and in the morphogenic process. The different tissues of the same plant and the same tissues at various developmental stages differ in their embryogenic response (Robert et al., 1986). Somatic embryos develop from single cells often produce genetically stable plants while multicellular origin in shoot morphogenesis results in the generation of genetically variable chimeric plants (Aly et al., 2002). Somatic embryogenesis cannot be defined as a specific response to one or more exogenously applied PGRs, instead, the formation of stressful conditions play a vital role in promoting dedifferentiation and induction of somatic embryo formation (Zavattieri et al., 2010). Studies at the molecular level support the correlation of PGRs to control the chromatin remodelling and gene expression during the induction of embryo and entire embryogenic process (Feher, 2008). The repeated induction of somatic embryo formation via secondary somatic embryogenesis holds great potential in mass plant propagation and the formation of artificial seeds in *C. sativum*. It is an easy and quick method of propagation, which offers the advantage of high multiplication rate, independent to the impact of explants and can be maintained for a long period by repeated cycles (Pavlovic et al., 2012). The development of resistance against biotic and abiotic stresses in tissue culture regenerated plants provides an alternative to conventional breeding having certain limitations including lack of useful variation and long time period (Bouquet and Torregrosa, 2003). Similarly the manipulation of physical or chemical mutagens, together with *in vitro* culture conditions, increased mutation frequency for disease tolerance in regenerated plants (Jain, 2010).

SYNTHETIC SEED TECHNOLOGY

Synthetic seeds are encapsulated regenerable plant materials including embryos, provides successful *in vitro* production of plantlets. The calcium-alginate coating provides protection against environmental stresses, furnishes nutritional micro-environment during short- or long-term storage promotes fast plantlet regeneration after storage and transport especially required for germplasm exchange and conservation (Palanyandy et al., 2015). The synthetic seed technology has various advantages including the low-cost production, higher potential to scale-up and easy handling between laboratories of national and international repute (Rai et al., 2009). The quality bead formation and germination frequency of synthetic seeds are influenced by sodium alginate and calcium chloride concentration and its exposure time. In *C. sativum* somatic embryos, encapsulated with 3% sodium alginate with 100 mM $CaCl_2$ produced uniform and moderately hard beads with optimum regeneration potential. Exposure time in $CaCl_2$ solution affects the germination ability of the synthetic seeds and decline in conversion may be due to growth inhibition caused by over absorption and penetration of calcium chloride during its long exposure time. The exposure time of 20 min in $CaCl_2$ solution resulted in the highest conversion rate in coriander which decreased with increasing exposure time.

The involvement of liquid MS in the preparation of gel matrix was found beneficial due to the presence of nutrients, synthetic seed gel matrix mimics endosperm of natural seed facilitates synthetic seed survival and germination (Mohanty and Das, 2013). We observed that the storage duration of synthetic seeds affects their viability and regeneration potential that decreases with storage time. During storage, continuous reduction in carbohydrate reserves reduces the regeneration potential of synthetic seeds (Ding et al., 1998). The storage of synthetic seeds at -20°C resulted in their poor conversion potential as well as the expiration of tissue viability. It is considered that the formation of ice crystals in tissues on exposure to low temperature, prevents the exchange of food material, induces frost injury/ stress and reduces the viability of tissue (Mittler, 2006). The synthetic seed regenerated plants were healthy and morphologically very similar to parent plants. The plant regeneration protocol developed from synthetic seeds (kept especially at low temperature for varied periods) will be very significant in the preservation of coriander germplasm for short to medium term basis. The combination of auxin and cytokinin favoured maturation and subsequent germination of somatic embryos. A similar observation of somatic embryo maturation on NAA and BA supplemented medium have been reported in other observations (Chand and Singh, 2001; Swamy et al., 2005).

PROTOPLAST TECHNOLOGY

Protoplasts are the cells from which cell walls have been deliberately removed and are surrounded by the cell membrane. After removal of the cell wall, protoplasts continue their pluripotentiality, a dedifferentiation process with simultaneous chromatin remodelling (Pulianmackal et al., 2014). Dedifferentiation is a developmental process which provides a progenitor cell source for the plant regeneration under *in vitro* conditions (Chupeau et al., 2013). The expression of cellular totipotency in protoplast culture is a complex developmental phenomenon that enables in studying cellular activities such as degradation and synthesis of the cell wall, differentiation, dedifferentiation, cellular communication and cell division (Yoo et al., 2007). We developed efficient protoplast isolation, culture and plant regeneration protocol via somatic embryogenesis in coriander (Figure 2). We used embryogenic cell suspension as a source of protoplast which appeared to be a very efficient source of regenerable protoplasts and produces dividing cells at a reasonable rate.

Embryogenic cell suspensions serve as a good source for the successful isolation of protoplasts in several investigated plant genera (Masani et al., 2013). The selection of genotypes, physiological status of donor plants/explants and culture conditions play an essential role in isolation and regeneration of protoplasts (Wiszniewska and Pindel, 2009). The protoplast isolation variability may be due to somaclonal variation within a genotype, or due to different antioxidant and phytochemical conditions that affect regeneration capacity (Pan et al., 2004). We observed that half-day incubation of embryogenic cell suspension in cell wall digestion enzyme solution was found to be optimum for isolation of viable protoplasts. Since protoplasts are fragile and sensitive to digesting enzymes, prolong incubation in enzyme solution led to protoplast breakage and dysfunction even though the yield may often high (Tudses et al., 2014). Such damage could be minimized by modifying the enzyme treatment or reducing time duration (Navratilova et al., 2000).

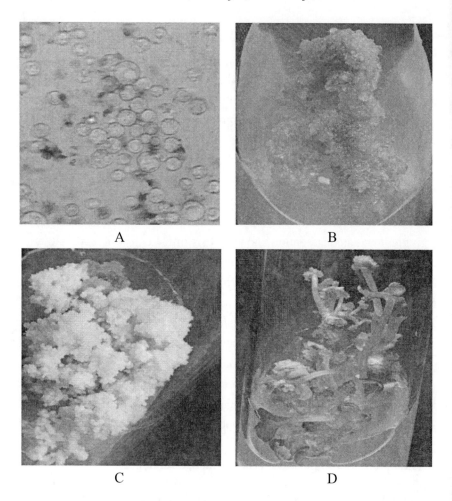

Figure 2. Isolated protoplasts and plant regeneration. (A) protoplasts; (B) protoplast derived callus; (C) somatic embryogenesis; (D) plant regeneration.

Culture media must be osmotically adjusted to prevent rupture of protoplasts during isolation and early culture due to the elimination of wall pressure (pressure potential)- a component of water potential (Duquenne et al., 2007). In protoplast culture media, sugars including mannitol, sorbitol and glucose are used as osmotic stabilizers and carbon sources with their concentrations are gradually reduced after cells start to synthesize a new wall and later are withdrawn at the time of macroscopic colony formation (Zhang et al., 2011). The first cell division often occurs within 2-3 days of culture and is affected by culture medium, osmoticum, plating density, storage conditions, and source tissue (Prange et al., 2010). In the conventional optimization process, one parameter is altered at a time keeping other variables constant, which enables one to assess the performance of specific treatment (Zhang et al., 2011). Addition of supplements, e.g., nurse cells into protoplast culture medium increases the plating efficiency (Karamian and Ranjbar, 2010). Navratilova et al., (2000) observed that the culture induced damage could be minimized by modifying enzyme treatments or reducing the exposure time with a solution. Chamani et al., (2012) also noted right enzyme treatment and incubation time are crucial in obtaining viable protoplasts while working on diverse tissues of different plant species. In coriander, we identified 2% cellulase, 1% pectinase and 0.2% macerozyme were the most appropriate enzyme combinations for the higher yield and viability of protoplasts. The same enzyme combination treatment was proved to be efficient in producing the maximum yield of protoplasts in several investigated plant genera (Maqsood et al. 2012).

Rastogi (2003) reported that increased hydrolytic enzyme level enhanced protoplasts number by facilitating forming more enzyme-substrate complexes and additional enzymes level was noted to be unresponsive in increasing protoplast yield. Higher enzyme level was noted to influence negatively on protoplast yield and viability as the enzymes pectinase and cellulase over-digested protoplasts (Raikar et al., 2008). Osmotic condition is also critically important and several osmotic compounds like mannitol, sorbitol, glucose or sucrose are commonly added to the enzyme mixtures and were noted to be very effective in a number of studied observations (Navratilova, 2004). In the present study, the use of 0.6M mannitol showed the highest viability (85.84%), significantly more than the use of 0.6 M sorbitol. Since mannitol had better influence, this osmotic element was identified as an ideal osmotic treatment in achieving maximum yield and viable protoplasts in coriander. Protoplasts, in general, are more stable in a slightly hypotonic environment as compared to the isotonic solution and appropriate osmotic bathing prevents protoplast from bursting and shrinking (Duquenne et al., 2011). These results showed that the use of individual enzymes was not effective in producing good protoplast yields. This is due to the substrate specificity of enzymes and also the enzyme activities are dependent on the substrate present in the tissue (Chen et al., 1995; Yeong et al., 2008).

The protoplast isolation and their regeneration depend on several factors including the genotype of the plant species, inoculation density and medium composition as well (Prange et al., 2010; Tiwari et al., 2010). Culture of isolated protoplasts was done in liquid MS at two different densities, i.e., 1×10^5 and 2×10^5 protoplasts ml^{-1}, while protoplasts divide at an optimal density of 2×10^5 protoplasts ml^{-1}. With the result, cell colonies were produced from individual protoplasts that grew steadily in culture. The density of protoplasts is therefore important, it influences cell division initiation as was reported earlier (Chamani and Tahami, 2016). The early phases of culture such as rapid cell wall synthesis and mitotic divisions are crucial in the successful development of protoplast-to-plant systems. Plating density of protoplasts has also an important role in potential cell division and subsequent colony formation (Davey et al., 2005). In most culture media the nutritional components used are insufficient to induce divisions when protoplasts are plated at low densely. We observed optimum protoplast division in a liquid medium, similar to several other studies (Tahami et al., 2014) where the division of protoplasts has been noted to be more vigorous in a liquid medium. In a solid medium, the release of accumulated toxic compounds from dying neighbouring protoplasts inhibits the division of other protoplasts. However in liquid medium accumulation of these toxic compounds may be prevented or diluted (Duquenne et al., 2007).

Davey et al., (2005) reported that the PGRs play a primary role in protoplast division and regeneration, but their concentrations and combinations require optimization during protoplast development. In our study, different PGRs like 2, 4-D, NAA and BA were used and tested for protoplast division, in which the 2, 4-D and BA combination showed the better results. The process of cell-wall regeneration, cell division and callus formation were expeditiously attained in PGR added medium. In NAA and BA added MS plating efficiency increased, similar to the study of Assani et al., (2001). Auxins play an important role in cell division and callus formation and are involved in the acquisition of somatic embryogenesis (Ma et al., 2011). The embryo development gets blocked in the presence of 2, 4-D, however, for further development and maturity, other PGRs treatments are involved in the optimization of maturation medium (Junaid et al., 2006). After maturation, another crucial step is the successful germination of somatic embryos in obtaining plantlets (Devi and Narmathabai, 2011). In our observation, germination of mature somatic embryos was fast in GA_3 and BA supplemented MS medium, similar to other studies (Ali et al., 2010). Nakano et al., (2005) confirmed enhanced germination of somatic embryos and development of plantlets in lower BA added MS medium. Germination of somatic embryos further increased via additional changes in the germination medium which matured under sub-optimal conditions (Stasolla and Yeung, 2003).

The somatic embryo development from protoplast-derived callus has been confirmed by several previous studies (Tomiczak et al., 2015). Additionally, Plant regeneration from protoplast-derived callus via organogenic pathway are reported in other studies (Rezazadeh and Niedz, 2015). In the regeneration process differentiation of root and shoot apical meristems are unique in embryo-derived plants and are often true-to-type (Stasolla and Yeung, 2003); here, the regenerated plants showed similarity in morphology with the parent plant. The protoplast isolation and plant regeneration protocol in coriander could be very useful in the formation of somatic hybrid in coriander.

The isolation of viable protoplasts and their successful regeneration into plants have made the plant genetic engineering easy and more result oriented (Burris et al., 2016). Protoplasts are important experimental systems in physiological, biochemical and genetic work (Pati et al., 2008). The design of an efficient regeneration protocol is important when implementing protoplast fusion to breeding programmes. Protoplast-based approach act as a useful tool to overcome limitations in conventional breeding including sexual incompatibility by enabling direct transfer of both nuclear and cytoplasm genome features into plant cells (Prange et al., 2012). In the generation of somatic hybrids with improved characteristics, protoplasts fusion of taxonomically divergent plants brings together nuclear and cytoplasmic genomes at interspecific and intergeneric levels to beat barriers in naturally occurring sexual incompatibility (Prange et al., 2012).

There are several methods employed for the successful fusion of protoplasts from different plant species or even different genera. The technique of electrofusion is preferred over chemical (PEG) mediated fusion by reducing membrane damage and organelle fusion and maintaining protoplast viability (Duquenne et al., 2007). Even though in some plants binary chemical fusion is more efficient, electrical fusion allows more efficient cell division and plant regeneration afterwards (Assani et al., 2005). The process of somatic fusion provides a promising tool for transfer of polygenic agronomical traits like resistances from wild to cultivated crop species. Bacterial wilt resistance has been successfully introduced from *Solanum chacoense* to *S. tuberosum* via protoplast fusion (Cai, 2003). Somatic hybrids formed between zinc accumulator *Thlaspi caerulescens* and *Brassica napus* have shown potential in bioremediation and environmental clean-up (Brewer et al., 1999). The resulting hybrid plants accumulated zinc and cadmium to concentrations normally toxic to *B. napus*, indicating that the transfer of trait for metal accumulation in plants is achievable by protoplast fusion. Somatic hybridisation and hybridization also offer the possibility of increasing genetic variability with the effective transfer of desired characters between plants in many crops (Liu et al., 2007). Direct gene transfer via protoplasts, on the other hand, is also an option for the introgression of agronomically useful genes, such as sterility and disease or insect resistance (Strauss et al., 1995). Protoplasts have an immense role in facilitating the genetic transformation of important plant species. Due to fluid mosaic characteristics of the plasma membrane, DNA uptake can be facilitated by the application of some chemical and/or physical methods. The treatments of PEG or electrofusion are the normal approaches used to induce DNA uptake into protoplasts. Protoplasts can also be co-transformed with more than one gene carried on the same or separate plasmids. Irradiation and heat shock treatment to recipient protoplasts increase the transformation frequency, probably by enhancing the recombination of genomic DNA with foreign DNA, or initiating repair mechanisms in favour of DNA integration (Grosser and Omar, 2010). DNA uptake into protoplasts has been especially important in transforming plants that are not amenable to other methods of gene delivery, particularly *Agrobacterium*-mediated transformation (Rakoczy-Trojanowska, 2002). Protoplast, as an experi-mental tool, has been used in plant cell proteomic research with the application of fluorescence-activated cell sorting by flow cytometry (Fukao et al., 2013). Protoplast based functional gene analysis by double-stranded RNA interference provides a rapid and cost-effective initial screening of genes of interest for further studies in plants such as *Arabidopsis thaliana*, *Coptis japonica*, rice, poplar etc. (Miki and Shimamoto, 2004; Dubouzet et al., 2005; Zhai et al., 2009).

SOMACLONAL VARIATIONS

The *in vitro* conditions inflict adaptive stress on the cells and can induce genetic variations in them. Sometimes the plants regenerated via callus stage exhibit genetic instability including sequence changes, polyploidization and aneuploidy (Kaeppler et al., 2000). However, the plants regenerated through somatic embryogenesis are often true to type, which demonstrate genetic stability and is an important consideration in the development of transgenic plants (Liu et al., 2013). In *C. sativum*, the long regenerative potential of induced callus without losing its ability is very crucial for several morphogenetic processes including genetic transformation. Under *in vitro*, callus/suspensions are exposed to stress conditions caused by medium constituents including plant growth regulators which may lead to mutations and generate somaclonal variants (Pal et al., 2007). These alterations may be either detrimental or a useful source of genetic variation for improvement in plant characteristics (Carvalho et al., 2004). In *C. sativum*, we observed genetic stability of *in vitro* regenerated plantlets, by using flow cytometry technique. The plants were regenerated through primary and secondary somatic embryogenesis by intervening callus phase, due to which the analysis of DNA content was essential. In regenerated C. *sativum*, the nuclear DNA content was unaltered and preserved even though the plants were obtained from a culture, added with the stressor (2, 4-D containing medium). The presence of 2, 4-D, NAA, and BA in the medium are considered to be the reason of genetic variability in culture (Clarindo et al., 2008). Although the presence of PGR is essential for embryogenic callus formation and genetic variations induced under *in vitro* culture conditions showed desired traits in a number of economically important plant species (Synman et al., 2011). The *in vitro* culture conditions in combination with mutation breeding using chemicals (Shah et al., 2009) and rays (Sharma et al., 2010) have been proved to increase mutation frequency in developing disease tolerant plants (Purwati and Sudarsono 2007; Jain 2010). Similarly, culture conditions induced transposable elements movement was also reported in developing variants in cultures (Kikuchi et al., 2003). It is thus necessary to assess the genetic variability of regenerated plants, induced through indirect somatic embryogenesis. There is increasing use of flow cytometry for the confirmation of genome size and ploidy changes in regenerated plants (Rewers et al., 2012). The flow cytometry-based investigation of nuclear DNA content is an efficient technique for analysing genome size and aneuploid/ploidy identification in many important plant genera (Sliwinska and Thiem, 2007).

The control of true-to-type is important during culture and for the synthesis of biologically active compounds (Chaturvedi et al., 2007). The genetic alteration may occur at the cellular or at the molecular level by involving a change in chromosome structure and numbers or at the nucleotide sequence level (Radic et al., 2005). The occurrence of these variations, however, needs assessment of genetic stability of regenerants. Currently, cytological studies are complemented by flow cytometry studies which together enable in studying cellular status and make an accurate assessment of any change in ploidy of *in vitro* regenerants (Rewers et al., 2012). Among the various techniques available in molecular biology, random amplified polymorphic DNA (RAPD) analysis is proved to be a reliable DNA marker for assessing genetic fidelity in many economically important regenerated plants (Valladares et al., 2006) and in essential cases when prior knowledge of plant genome is unavailable (Agarwal et al., 2008). The assessment of genetic variation induced under *in vitro* culture conditions using chromosomal analysis, flow cytometry, and DNA fingerprinting are also used to ensure true-to-type regeneration process (Jain, 2001). The confirmation of genetic fidelity is a valuable tool to preserve desirable properties of *in vitro* grown plants and germplasm lines (Thiem et al., 2013). The occurrence of somaclonal variations has been reported from protoplast-derived regenerated plants (Tomiczak et al., 2015). These variations include changes in leaf and flower morphology, fertility, improvement in disease resistance and secondary metabolite yield, while the most noticeable change of somaclonal variation is the regeneration of polyploidy through endopolyploidization or spontaneous cell fusion (Nwauzoma and Jaja, 2013).

ESSENTIAL OIL AND ELICITATION

Essential oils are complex mixtures of low molecular weight lipophilic compounds like terpenes, terpenoids and phenolics produced in cytoplasm and plastids of plant cells via malonic acid, mevalonic acid and methyl-D-erythritol-4-phosphate pathways (Voon et al., 2012). Essential oils may comprise about 100 different plant secondary metabolites belonging to a variety of chemical classes (Carson and Hammer, 2011). However, the presence of a mixture of molecules modifies the activity to exert significant effect (Isman et al., 2008). The lipophilic nature of essential oils appears to play a crucial role in their antimicrobial activity by disrupting membrane permeability and osmotic balance of the cell. Terpenes are hydrocarbon with several isoprene (C_5H_8) units while the terpenoids are a biochemical modification of terpenes via enzymes that add oxygen molecules or remove methyl group (Caballero et al., 2003). In general, terpenoids are more antimicrobial than terpenes (Burt, 2004). Plants producing essential oils belong to various genera distributed to around 60 families, some families such as Alliaceae, Apiaceae, Asteraceae, Lamiaceae, Myrtaceae, Poaceae and Rutaceae are well known for their ability to produce essential oils of medicinal and industrial value (Vigan, 2010; Hammer and Carson, 2011). All the essential oil producing plant families are rich in terpenoids while plant families like Apiaceae (Umbelliferae), Lamiaceae, Myrtaceae,

Piperaceae and Rutaceae contain phenylpropanoids more frequently (Chami et al., 2004). Plants of these families are exploited for essential oil production at a commercial level for example; coriander, anise, dill and fennel oils are extracted from *Coriandrum sativum*, *Pimpinella anisum*, *Anethum graveolens* and *Foeniculum vulgare* plants respectively belonging to the family Apiaceae and are known for their antifungal, antibacterial, anticancer and antiviral activities (Raut and Karuppayil, 2014). Essential oils play an important role in human life having multiple applications in medicines, cosmetics, food processing etc. (Yentema et al., 2007). Plants producing essential oils are the primary constituent of fragrance and aroma; essential oils are used in food and pharmaceutical industries owing to various medicinal (therapeutic, antimicrobial and antioxidant and other) activities (Hussain et al., 2008; Teixeira et al., 2013). These compounds are also used as herbicides, pesticides and anticancer compounds (Mahmoud and Croteau, 2002; Abraham et al., 2003; Burfield and Reekie, 2005). The essential oils have a role in plant defence and pollinator attraction among other ecological functions these compounds confer to plants' fitness under environmental variation, results in quantitative and qualitative product variation (Taiz and Zeiger, 2004). Essential oils are considered eco-friendly and are generally exempted from toxicity data requirements by the Environment Protection Agency (Prakash et al., 2010). The essential oils can easily be released as a botanical fumigant to the acceptable agri-commodity items with the help of modern encapsulation technology. The use in agri-commodity items are economically viable, effective and it leaves minimal residues in applicable food items (Isman, 2006). The use of synthetic antioxidants like butylated hydroxyanisole and butylated hydroxytoluene in the food industry as inhibitors of lipid peroxidation has caused many problems due to their volatile nature and instability at high temperatures (Dapkevicius et al., 1998). Many plant species that are being used as a tea, fruits, spices and vegetables have been attractive to scientists as natural sources of antioxidant compounds (Paur et al., 2011; Hwang et al., 2014). These plants have been screened for their antioxidant capacities and are safer than synthetic ones that led to their introduction as natural antioxidants (Schuler, 1990). Antioxidants function in several ways and important one is related to the protection of cells against oxidative stress, caused by reactive oxygen species (ROS), formed during normal cell functioning, and damages cells linked with cancers and other cardiovascular diseases (Koleva et al., 2002; Ou et al., 2002). Essential oils or some related components are used in perfumes and make-up products, sanitary products, dentistry, agriculture, as food preservatives, additives and as natural remedies (Bakkali et al., 2008). In recent years, due to renewed interest in natural products, plant essential oils receive more focus on phytomedicine (Zu et al., 2010; Sylvestre et al., 2006). It is also essential to develop a better basic research understanding especially in the area of antioxidant, antimicrobial and potential anticancer activities of essential oils (Mimica-Dukic et al., 2004; Sylvestre et al., 2005). The development of multiple drug/chemical resistance both in humans and plant pathogenic microorganisms due to overuse of commercial antimicrobial drugs/chemicals in the treatment of infectious diseases have forced the scientific community to search alternative and safe substances from various sources including medicinal plants (Cordell, 2000). The efficiency of essential oils in inhibiting pathogen growth and their ability to increase shelf life of food products compelled the scientific community to explore the biological activity of essential oils more and more for commercial viewpoint (Prakash et al., 2011). In future, the essential oils may play a key role in integrated pest management programmes and may provide immense opportunity to agriculture sectors to develop plant-based compounds against pest and pesticide originated issues. There are several reports available on essential oil from *in vivo* grown plants but only a few studies such as in *Cananga odorata* (Lindain et al., 2008), *Foeniculum vulgare* (Khodadadi et al., 2013) etc. have been conducted for the production of essential oil using *in vitro* cultures. The essential oil yield and composition depends on various factors including plant genetics, developmental stage, application of fertilizers, geographic location, surrounding climate, stress during growth or maturity, the post-harvest drying, storage and the method of extraction (Hussain et al., 2008).

Natural flora plays an important role in our life as intact the biochemical factory for the production of both primary metabolites like sugars, amino acids and fatty acids as well as secondary metabolites including alkaloids, flavonoids, volatile oils with pharmaceutical significance (Namdeo, 2007). They are usually found in low concentra-tions under *in vivo* conditions that makes extraction from natural vegetation difficult and expensive (Zabala et al., 2010). Under *in vivo*, secondary metabolites are produced in a very low quantity depending on the physiological and developmental stage of the plant and are important to plants in their interactions with the environment (Oksman-Caldentey and Inze, 2004). The plant cells possess biosynthetic totipotency which allows them to produce the same chemical compounds *in vitro* as *in vivo* (Tamer and Mavituna, 1996). However, plant cell cultures represent a potential source of many such compounds that cannot be produced by microbial cells or chemical syntheses (Di-Cosmo and Misawa, 1995). *In vitro* cell cultures offer an alternative technique for the production of these metabolites which can further be enhanced by the application of elicitors (Roat and Ramawat, 2009). Cell cultures of several plants do not produce sufficient amount of secondary metabolites (Rao and Ravishankar, 2002), but the yield can be increased by the intervention of elicitors such as methyl jasmonate (MeJA), salicylic acid, chitosan and heavy metals (Poulev et al., 2003). Efforts have been made on optimizing cultural conditions, employing precursor feeding and on the use of elicitation techniques to obtain the enriched level of bioactive compounds (Ahmad et al., 2013). Elicitation of plant cell cultures is considered as the most efficient strategy to increase the secondary metabolite accumulation including essential oil in many plant species (Ribkahwati et al., 2015). The process of elicitation leads to enhanced production of secondary metabolites in plants for protection against various pathogens, herbivores and to survive under stress conditions (Zhao et al., 2005). Methyl jasmonate (MeJA) has been used as an elicitor for enhancing the production of secondary compounds in plant cell cultures of *Panax ginseng* for obtaining ginsenoside (Thanh et al., 2005), anthocyanin in *Melastomam alabathricum* (See et al., 2011), in hairy roots cultures for glycyrrhizin in *Glycyrrhiza inflata* (Wongwicha et al., 2011), diterpenoid in *Salvia sclarea* (Kuzma et al., 2009). JA and its methyl ester MeJA are considered important signalling compounds in the elicitation process leading to accumulation of various secondary metabolites (Shimizu et al., 2010). These also play an important role as cellular regulators in various developmental processes, such as seed germination, fertility, fruit ripening and senescence (Wasternack and Hause, 2002).

In *C. sativum* the effect of MeJA on essential oil yield was studied on *in vitro* grown cultures. The yield of essential oil was improved on the addition of MeJA. Similar to other reports of MeJA induced synthesis of terpenoids (Ribkahwati et al., 2015), we noted that the total amount of essential oil significantly increased with increase in the concentration of MeJA in the culture medium. In *C. sativum*, however, the callus tissue showed the absence of essential oil but started to accumulate with the differentiation of tissues and maximum at mature stages of the somatic embryo. The effect of MeJA is based on duration and concentration of elicitation. We observed the higher level of MeJA (300µM) was efficient in stimulating essential oil synthesis compared to lower levels. The accumulation of valuable bioactive compounds has been successfully promoted by a diverse group of compounds. The mechanism of elicitation is, however, diverse in plants and the elicitation mechanism includes the involvement of studies at various levels viz. cell wall, cytosolic messenger molecules and biosynthetic pathway. Radman et al. (2003) suggested the occurrence of 'elicitor-receptor interaction' is the bases for a rapid array of biochemical responses in most cases of elicitation. It is worth mentioning that the elicitor is able to trigger the synthesis of novel compounds; hence, this study is of fundamental and applied importance. Our results suggest that MeJA acted as an inducer in enhancing the biosynthesis of essential oils in *C. sativum*. Elicitors bind to a receptor protein present in the plasma membrane and activate specific genes through signal transduction pathways as a part of defence response and promote the biosynthesis of secondary metabolites in elevated amounts (Mishra et al., 2012). The development of *in vitro* cultures is an alternative approach to the intact plant for secondary metabolite synthesis. However recent approaches also involve the application of molecular biology techniques to enhance the metabolic pathways of specific compounds. Many factors influence the elicitation and production of bioactive compounds such as elicitor concentration, duration and age of subculture etc. (Patel and Krishnamurthy, 2013). The secondary metabolites start accumulating when *in vitro* growth slows down, indicates the existence of an inverse relationship between growth and secondary metabolite production (Farzami-Sepehr and Ghorbanli, 2002). The concentrations of elicitor strongly affect the intensity of secondary metabolite synthesis that varies according to the plant species (Ramakrishna and Ravishankar, 2011). Future studies involving the role of genes/enzymes in regulating MeJA- mediated phenylpropanoid synthesis may shed some light in this direction.

CONCLUSION

In conclusion, a better understanding of the specific requirements at each critical step of the regeneration process is needed in order to improve their entire regenerative performance. The occurrence of both primary and secondary somatic embryogenesis with genetic stability of in vitro-raised plants in *Coriandrum sativum* possesses the potential for large-scale application in plant biotechnology. The formation of synthetic seeds and their regeneration after storage may provide alternative propagation method as well as enable conservation possibility of elite important coriander germplasm. The development of efficient protoplast-isolation protocol taking into account different concentrations and combinations of enzyme and osmoticum for maximum yield and viability are key steps in protoplast fusion and somatic hybridization to be employed in closely and widely related coriander genotypes. It opens an opportunity for improvement in the plant through a transgenic approach. Moreover, cell/tissue cultures are an efficient source for secondary metabolite production.

REFERENCES

Abrahim, D., Francischini, A.C., Pergo, E.M., Kelmer-Bracht, A.M. and Ishii-Iwamoto, E. (2003).Effects of α-pinene on the mitochondrial respiration of maize seedlings. *Plant Physiology and Biochemistry*, 41, 985-991.

Agarwal, M., Shrivastava, N. and Padh, H. (2008). Advances in molecular marker techniques and their applications in plant science. *Plant Cell Reports*, 27, 617-631.

Ahmad, S., Garg, M., Tamboli, E.T., Abdin, M.Z. and Ansari, S.H. (2013). In vitro production of alkaloids: Factors, approaches, challenges and prospects. *Pharmacognosy Reviews*, 7(13), 27-33.

Ahsan, M.Z., Majidano, M.S., Channa, A.R., Panhwar, F.H., Soomro, A.W., Khaskheli, F.I. and Rashid, K. (2014). Regeneration of cotton (*Gossypium hirsutum* L.) through asexual Methods, a review. *American-Eurasian Journal of Agricultural and Environmental Science*, 14(12), 1478-1486.

Ali, S.B.G.M., Kourosh, V., Hassan, B.S., Siamak, K. and Charles, L. (2010). Enhancement of maturation and germination of somatic embryos in Persian walnut (*Juglans regia* l.) using osmolites, hormones and cold treatments. *African Journal of Food Science*, 4(12), 735-743.

Aly, M.A. M., Rathinasabapathi, B. and Kelly, K. (2002). Somatic embryogenesis in perennial statice *Limonium bellidifolium* Plumbaginaceae. *Plant Cell, Tissue and Organ Culture*, 68, 127-135.

Assani, A., Chabane, D., Haıcour, R., Bakry, F., Wenze, G. and Foroughi-Wehr, B. (2005). Protoplast fusion in banana (*Musa* spp.): comparison of chemical (PEG: polyethylene glycol) and electrical procedure. *Plant Cell, Tissue and Organ Culture*, 83, 145-151.

Assani, A., Haicour, R., Wenzel, G., Cote, F., Bakry, F., Foroughi-Wehr, B., Ducreux, G., Aguillar, M.E. and Grapin, A. (2001). Plant regeneration from protoplasts of dessert banana cv. grande naine (*Musa* spp., Cavendish Sub-group AAA) via somatic embryogenesis. *Plant Cell, Reports*, 2, 482-488.

Bakkali, F., Averbeck, S., Averbeck, D. and Idaomar, M. (2008). Biological effects of essential oils - a review. *Food and Chemical Toxicology*, 46, 446-475.

Bandoni, A.L., Mizrahi, I. and Juarez, M.A. (1998).Composition and quality of essential oil of coriander (*Coriandrum sativum* L.) from Argentina. *Journal of Essential Oil Research*, 10, 581-584.

Bhuiyan, M.N. I., Begum, J. and Sultana, M. (2009). Chemical composition of leaf and seed essential oil of *Coriandrum sativum* L. from Bangladesh. *Bangladesh Journal of Pharmacology*, 4, 150-153.

Bouquet, A. and Torregrosa, L. (2003). Micropropagation of the grapevine (*Vitis* spp.). In: Jain, S.M. and Ishii, K. (Eds.), *Micropropagation of woody trees and fruits* (pp. 319-352). The Netherlands: Kluwer Academic Publishers, Dordrecht.

Brewer, E.P., Saunders, J.A., Angle, J.S., Chaney, R.L. and McIntosh, M.S. (1999). Somatic hybridization between the zinc accumulator *Thlaspi caerulescens* and *Brassica napus*. *Theoretical and Applied Genetics*, 99, 761-771.

Burfield, T. and Reekie, S.L. (2005). Mosquitoes, malaria and essential oils. *International Journal of Aromatherapy*, 15, 30-41.

Burris, K.P., Dlugosz, E.M., Collins, A.G., Stewart-Jr C.N. and Lenaghan, S.C. (2016). Development of a rapid, low-cost protoplast transfection system for switchgrass (*Panicum virgatum* L.). *Plant Cell Reports*, 35, 693-704.

Burt, S. (2004). Essential oils: their antibacterial properties and potential applications in foods- a review. *International Journal of Food Microbiology*, 94, 223-253.

Caballero, B., Trugo, L.C. and Finglas, P.M. (2003). *Encyclopedia of Food Sciences and Nutrition (Second edition)*. Academic Press, Amsterdam.

Cai, X. (2003). *Creation and genetic analysis of new germplasms resistant to Ralstonia solanacearum via cell fusion in potatoes*. Dissertation, Huazhong Agricultural University.

Carson, C.F. and Hammer, K.A. (2011). Chemistry and bioactivity of essential oils. In: Thormar, H. (Ed.), *Lipids and essential oils as antimicrobial agents* (pp. 203-238). UK, John Wiley and Sons.

Carvalho, L.C., Goulao, L., Oliveira, C., Goncalves, J.C. and Amancio, S. (2004). RAPD assessment for identification of clonal identity and genetic stability of in vitro propagated Chestnut hybrids. *Plant Cell, Tissue and Organ Cult*, 77, 23-27.

Chamani, E. and Tahami, S.K. (2016). Efficient protocol for protoplast isolation and plant regeneration of *Fritillaria imperialis* L. *Journal of Agricultural Science and Technology*, 18, 467-482.

Chamani, E., Tahami, S.K., Zare, N., Zakaria, R.A., Mohebodini, M. and Joyce, D. (2012). Effect of different cellulase and pectinase enzyme treatments on protoplast isolation and viability in *Lilium ledebeourii* Bioss. *Notulae Botanicae Horti Agrobotanici Cluj-Napoca* 40(2), 123-128.

Chami, F., Chami, N., Bennis, S., Trouillas, J. and Remmal, A. (2004). Evaluation of carvacrol and eugenol asprophylaxis and treatment of vaginal candidiasis in an immune suppressed ratmodel. *Journal of Antimicrobial Chemotherapy*, 54, 909-914.

Chand, S. and Singh, A.K. (2001). Dirrect somatic embryogenesis from zygotic embryos of a timber yielding leguminous tree, *Hardwickia binata* Roxb. *Current Science*, 80, 882-888.

Chaturvedi, H.C., Jain, M. and Kidwai, N.R. (2007). Cloning of medicinal plants through tissue culture- A review. *Indian Journal of Experimental Biology*, 45, 937-948.

Chen, R.R., Zhang, J.T., Ping, L.B., Guo, S.S., Hao, J.P. and Zhou, X.M. (1995). Somatic embryogenesis and artificial seed in coriander (*Coriandrum sativum* L.). *Biotechnology in Agriculture and Forestry*, 31, 334-342.

Chithra, V. and Leelamma, S. (1997). Hypolipidemic effect of coriander seeds (*Coriandrum sativum*): mechanism of action. *Plant Foods for Human Nutrition*, 51 (2), 167-172.

Chithra, V. and Leelamma, S. (2000). *Coriandrum sativum*- effect on lipid metabolism in 1, 2- dimethyl hydrazine induced colon cancer. *Journal of Ethnopharmacology*, 71, 457-463.

Chupeau, M.C., Granier, F., Pichon, O., Renou, J.P., Gaudin, V. and Chupeaua, Y. (2013). Characterization of the early events leading to totipotency in an *Arabidopsis* protoplast liquid culture by temporal transcript profiling. *Plant Cell,* 25(7), 2444-2463.

Clarindo, W.R., Carvalho, C.R., Arauojo, F.S., Abreu, I.S. and Otoni, W.C. (2008). Recovering polyploid papaya in vitro regenerants as screened by flow cytometry. *Plant Cell, Tissue and Organ Culture,* 92, 207-214.

Cordell, G.A. (2000). Biodiversity and drug discovery a symbiotic relationship. *Phytochemistry,* 55, 463-480.

Cortes, E.J., Gomez, A.S. and Villalobos, P.R. (2004). Antimutagenicity of coriander (*Coriandrum sativum* L.) juice on the mutagenesis produced by plant metabolites of aromatic amines. *Toxicology Letters,* 153, 283-292.

Coskuner, Y. and Karababa, E. (2007). Physical properties of coriander seed (*Coriandrum sativum* L.). *Journal of Food Engineering,* 80(2), 408-416.

Dapkevicius, A., Venskutonis, R., van Beek, T.A. and Linssen, P.H. (1998). Antioxidant activity of extracts obtained by different isolation procedures from some aromatic herbs grown in Lithuania. *Journal of Science Food Agriculture,* 77, 140-146.

Darughe, F., Barzegar, M. and Sahari, M.A. (2012). Antioxidant and antifungal activity of Coriander (*Coriandrum sativum* L.) essential oil in cake. *International Food Research Journal,* 19 (3), 1253-1260.

Davey, M.R., Anthony, P., Power, J.B. and Lowe, K.C. (2005). Plant protoplasts: status and biotechnological perspectives. *Biotechnology Advances,* 23, 131-171.

Delporte, F., Pretova, A., Jardin, P. and Watillon, B. (2014). Morpho-histology and genotype dependence of in vitro morphogenesis in mature embryo cultures of wheat. *Protoplasma,* 251, 1455-1470.

Devi, B.C. and Narmathabai, V. (2011). Somatic embryogenesis in the medicinal legume *Desmodium motorium* (Houtt.)Merr. *Plant Cell, Tissue and Organ Culture,* 106, 409-418.

Di-Cosmo, F. and Misawa, M. (1995). Plant cell and tissue culture: alternatives for metabolite production. *Biotechnology Advances,* 13(3), 425-453.

Diederichsen, A. (1996). Results of a characterization of a germplasm collection of coriander (*Coriandrum sativum* L.) in *the Gatersleben genebank* (pp. 45-48). Inter. Symp. Breeding Res. Med. Aromatic Plants, Quedlinburg, Germany.

Ding, C.K., Chachin, K., Hamauzi, Y., Ueda, Y. and Imahori, Y. (1998). Effect of storage temperature and physiology and quality of loquot fruit. *Postharvest Biology and Technology*, 14, 300-315.

Dubouzet, J.G., Morishige, T., Fujii, N., An, C.I., Fukusaki, E., Ifuku, K. and Sato, F. (2005). Transient RNA silencing of scoulerine 9-O-methyltransferase expression by double stranded RNA in *Coptis japonica* protoplasts. *Bioscience Biotechnology and Biochemistry*, 69, 63-70.

Duquenne, B., Eeckhaut, T., Werbrouck, S. and Huylenbroeck, J.V. (2007). Effect of enzyme concentrations on protoplast isolation and protoplast culture of *Spathiphyllum* and *Anthurium*. *Plant Cell Tissue and Organ Culture*, 91, 165-173.

Edwards, R. (2004). No remedy in sight for herbal ransack. *New Scientist*, 181, 10-11.

Eikani, M.H., Golmohammad, F. and Rowshanzamir, S. (2007). Supercritical water extraction of essential oils from coriander seeds (*Corinadrum sativum* L.). *Journal of Food Engineering*, 80(2), 735-740.

El-Dougdoug, K.A. and El-Shamy, M.M. (2011). Management of viral diseases in banana using certified and virus tested plant material. *African Journal of Microbiology Research*, 5(32), 5923-5932.

El-Meskaoui, A. (2013). Plant cell tissue and organ culture biotechnology and its application in medicinal and aromatic plants. *Medicinal and Aromatic Plants*, 2(3), doi:10.4172/2167-0412.1000e147.

Emamghoreishi, M., Khasaki, M. and Aazam, M.F. (2005). *Coriandrum sativum* L.: evaluation of its anxiolytic effects in the elevated plus maze. *Journal of Ethnopharmacology*, 96, 365-370.

Farzami-Sepehr, M. and Ghorbanli, M. (2002). Effects of nutritional factors on the formation of Anthraquinones in callus cultures of *Rheum ribes*. *Plant Cell, Tissue and Organ Culture*, 68, 171-175.

Feher, A. (2008). The initiation phase of somatic embryogenesis: what we know and what we don't. *Acta Biologica Szegediensis*, 52(1), 53-56.

Feher, A., Pasternak, T.P. and Dudits, D. (2003). Transition of somatic plant cells to an embryogenic state. *Plant Cell, Tissue and Organ Culture*, 74, 201-228.

Fuente, E.B., Gil, A., Lenardis, A.E., Pereira, M.L., Suarez, S.A., Ghersa, C.M. and Grass, M.Y. (2003). Response of winter crops differing in grain yield and essential oil production to some agronomic practices and environmental gradient in the Rolling Pampa, Argentina. *Agric. Ecosys. Environ*, 99, 159-169.

Fukao, Y.1., Yoshida, M., Kurata, R., Kobayashi, M., Nakanishi, M., Fujiwara, M., Nakajima, K. and Ferjani, A. (2013). Peptide separation methodologies for in depth proteomics in Arabidopsis. *Plant Cell Physiology*, 54(5), 808-815.

Gray, A.M. and Flatt, P.R. (1999). Insulin-releasing and insulin-like activity of the traditional anti-diabetic plant *Coriandrum sativum* (coriander). *Brazilian Journal of Nutrition,* 81 (3), 203-209.

Grosser, J.W. and Omar, A.A. (2010). Protoplasts- An increasingly valuable tool in plant research. In: Trigiano, R.N. and Gray, D.J. (Eds.), *Plant tissue culture, development and biotechnology* (pp. 349-363). CRC Press.

Guo, Y., Bai, J. and Zhang, Z. (2007). Plant regeneration from embryogenic suspension-derived protoplasts of ginger (*Zingiber officinale* Rosc.). *Plant Cell, Tissue and Organ Culture*, 89, 151-157.

Hakkinen, S.T. and Ritala, A. (2010). Medicinal compounds produced in plant cell factories. In: Arora, R. (Ed.) Medicinal plant biotechnology (pp. 13-35). CABI, Oxfordshire, UK.

Hammer, K.A. and Carson, C.F. (2011). Antibacterial and antifungal activities of essential oils. In: Thormar, H. (Ed.), *Lipids and essential oils as antimicrobial agents* (pp. 255-306). John Wiley and Sons, Ltd, UK.

Helle, W., Anne, B.S. and Karl, E.M. (2004). Antioxidant activity in extracts from coriander. *Food Chemistry*, 88(2), 293-297.

Hussain, A., Qarshi, I.A., Nazir, H. and Ullah, I. (2012a). Plant tissue culture: Current status and opportunities. In: Leva, A. and Rinaldi, L.M.R. (Eds.), *Recent advances in plant in vitro culture* (pp. 1-28). UK, InTech.

Hussain, A.I., Anwar, F., Hussain Sherazi, S.T. and Przybylski, R. (2008). Chemical composition, antioxidant and antimicrobial activities of basil

(*Ocimum basilicum*) essential oils depends on seasonal variations. *Food Chemistry*, 108, 986-995.

Hussain, M.S., Fareed, S., Ansari, S., Rahman, M.A., Ahmad, I.Z. and Saeed, M. (2012b). Current approaches toward production of secondary plant metabolites. *Journal of Pharmacology and Bioallied Science*, 4(1), 10-20.

Hwang, E., Lee, D.G., Park, S.H., Oh, M.S. and Kim, S.Y. (2014). Coriander leaf extract exerts antioxidant activity and protects against UVB-induced photoaging of skin by regulation of procollagen type I and MMP-1 expression. *Journal of Medicinal Food*, 17(9), 985-95.

Innocent, B.X., Fathima, M.S. A. and Dhanalakshmi(2011). Studies on the immouostimulant activity of *Coriandrum sativum* and resistance to *Aeromonas hydrophila* in Catla catla. *Journal of Applied Pharmaceutical Science*, 1(7), 132-135.

Isman, M.B. (2006). Botanical insecticides, deterrents and repellents in modern agriculture and an increasingly regulated world. *Annuals of Review in Entomology*, 51, 45-66.

Isman, M.B., Wilson, J.A. and Bradbury, R. (2008). Insecticidal activities of commercial rosemary oils (*Rosmarinus officinalis*) against larvae of *Pseudaletia unipuncta* and *Trichoplusia ni* in relation to their chemical compositions. *Pharmaceutical Biology*, 46, 82-87.

Jain, S.M. (2001). Tissue culture-derived variation in crop improvement. *Euphytica*, 118, 153-166.

Jain, S.M. (2010). In vitro mutagenesis in banana (*Musa* spp.) improvement. *Acta Hortculturae*, 879, 605-614.

Ji, H.F., Li, X.J. and Zhang, H.Y. (2009). Natural products and drug discovery. Can thousands of years of ancient medical knowledge lead us to new and powerful drug combinations in the fight against cancer and dementia. *EMBO Reports*, 10(3), 194-200.

Jimenez, V.M. and Thomas, C. (2005). Participation of plant hormones in determination and progression of somatic embryogenesis. In: Mujib, A. and Samaj, J. (Eds.), *Somatic Embryogenesis: Plant Cell Monographs* (pp. 103-118). Springer-Verlag, Berlin/Heidelberg.

Junaid, A., Mujib, A., Bhat, M.A. and Sharma, M.P. (2006). Somatic embryo proliferation, maturation and germination in *Catharanthus roseus*. *Plant Cell, Tissue and Organ Culture*, 84, 325-332.

Kaeppler, S.M., Kaeppler, H.F. and Rhee, Y. (2000). Epigenetic aspects of somaclonal variation in plants. *Plant Molecular Biology*, 43, 179-188.

Kala, C.P., Dhyani, P.P. and Sajwan, B.S. (2006). Developing the medicinal plants sector in northern India: challenges and opportunities. *Journal of Ethnobiology and Ethnomedicine*, 2(32), 1-15.

Karamian, R. and Ranjbar, M. (2010). Somatic embryogenesis and plantlet regeneration from protoplast culture of *Muscari neglectum* Guss. *African Journal of Biotechnology*, 10, 4602- 4607.

Khan, T.A., Mazid, M. and Mohammad, F. (2011). Status of secondary plant products under abiotic stress: an overview. *Journal of Stress Physiology and Biochemistry*, 7, 75-98.

Khodadadi, E., Aharizad, S., Mohammadi, S.A., khodadadi, E., Kosarinasab, M. and Sabzi, M. (2013). Chemical composition of essential oil compounds from the callus of fennel (*Foeniculum vulgare* Miller.). *International Journal of Agronomy and Agricultural Research*, 3(11), 1-6.

Kikuchi, K., Terauchi, K., Wada, M. and Hirano, H.Y. (2003). The plant MITE mPing is mobilized in anther culture. *Nature,* 421, 167-170.

Koleva, I.I., Van-Beek, T.A., Linssen, J.P. H.; De-Groot, A. and Evstatieva, L.N. (2002). Screening of plant extracts for antioxidant activity: a comparative study on three testing methods. *Phytochemical Analysis*, 13(1), 8-17.

Kumar, S.R.S., Krishna, V., Pradeepa, V.K., Kumar, K.G. and Gnanesh, A.U. (2012). Dirrect and indirrect method of plant regeneration rom root explants of *Caesalpinia bonduc* (L.) Roxb.- A threatned medicinal plant of weatern ghats. *Indian Journal of Experimental Biology*, 50, 910-917.

Kuzma, L., Bruchajzer, E. and Wysokinska, H. (2009). Methyl jasmonate effect on diterpenoid accumulation in *Salvia sclarea* hairy root culture in shake flasks and sprinkle bioreactor. *Enzyme and Microbial Technology*, 44, 406-410.

Lal, A.A., Kumar, T., Murthy, P.B. and Pillai, K.S. (2004). Hypolipidemic effect of *Coriandrum sativum* L. in triton-induced hyperlipidemic rats. *Indian Journal of Experimental Biology*, 42(9), 909-912.

Lawrence, B.M. (1992). A planning scheme to evaluate new aromatic plants for the flavor and fragrance industries. In: Janick, J. and Simon, J.E. (Eds.), *Proceedings of the second national symposium. New Crops: exploration, research, and commercialization* (pp 620-627). New York, John Wiley and Sons, Inc.

Lindain, A.F., Reglos, R.A., de-Guzman, C.C. and Cedo, M.C. (2008). Tissue culture and essential oil production from callus cultures of Ilang-Ilang (*Cananga odorata* (Lamk) Hook. f. and Thomson). *Philippine Agricultural Scientist*, 91(3), 251-260.

Liu, F., Ryschka, U., Marthe, F., Klocke, E., Schumann, G. and Zhao, H. (2007). Culture and fusion of pollen protoplasts of *Brassica oleracea* L. var. italica with haploid mesophyll protoplasts of *B. rapa* L. ssp. pekinensis. *Protoplasma*, 231, 89-97.

Liu, W., Liang, Z., Shan, C., Marsolais, F. and Tian, L. (2013). Genetic transformation and full recovery of alfalfa plants via secondary somatic embryogenesis. *In Vitro Cell Developmental Biology- Plant*, 49, 17-23.

Ma, G., Lu, J., da Silva, J.A.T., Zhang, X. and Zhao, J. (2011). Shoot organogenesis and somatic embryogenesis from leaf and shoot explants of *Ochnainte gerrima* (Lour). *Plant Cell, Tissue and Organ Culture*, 104, 157-162.

Mahmoud, S.S. and Croteau, R.B. (2002). Strategies for transgenic manipulation of monoterpene biosynthesis in plants - Review. *Trends in Plant Science*, 7(8), 366-373.

Maqsood, M., Mujib, A. and Zahid, H.S. (2012). Synthetic seed development and conversion to plantlet in *Catharanthus roseus* (L.) G. Don. *Biotechnology*, 11(1), 37-43.

Masani, M.Y.A., Noll, G., Parveez, G.K.A., Sambanthamurthia, R. and Prufer, D. (2013). Regeneration of viable oil palm plants from protoplasts by optimizing media components, growth regulators and cultivation procedures. *Plant Science*, 210, 118-127.

Matasyoh, J.C., Maiyo, Z.C., Ngure, R.M. and Chepkorir, R. (2009). Chemical composition and antimicrobial activity of the essential oil of *Coriandrum sativum* L. *Food Chemistry*, 113, 526-529.

Medrano, R.M.E., Maldonado-Borges, J.I., Burgos-Tan, M.J., Valadez-Gonzalez, N. and Ku-Cauich, J.R. (2014). Using flow cytometry and cytological analyses to assess the genetic stability of somatic embryo-derived plantlets from embryogenic *Musa acuminata* Colla (AA) ssp. *malaccensis* cell suspension cultures. *Plant Cell, Tissue and Organ Culture*, 116, 175-185.

Micke, O., Hubner, J. and Munstedt, K. (2009). Ayurveda. *Der Onkologe*, 15, 792-798.

Miki, D. and Shimamoto, K. (2004). Simple RNAi vectors for stable and transient suppression of gene function in rice. *Plant Cell Physiology*, 45, 490-495.

Mimica-Dukic, N., Bozin, B., Sokovic, M. and Simin, N. (2004). Antimicrobial and antioxidant activities of *Melissa officinalis* L. (Lamiaceae) essential oil. *Journal of Agricultural Food Chemistry*, 52, 2485-2489.

Mishra, A.K., Sharmab, K. and Misra, R.S. (2012). Elicitor recognition, signal transduction and induced resistance in plants. *Journal of Plant Interactions*, 7(2), 95-120.

Mittler, R. (2006). Abiotic stress, the field environment and stress combination. *Trends in Plant Science*, 11, 15-19.

Mohanty, P. and Das, J. (2013). Synthetic seed technology for short term conservation of medicinal orchid *Dendrobium densiflorum* Lindl. Ex Wall and assessment of genetic fidelity of regenerants. *Plant Growth Regulation*, 70, 297-303.

Nakano, M., Tanaka, S., Kagami, S. and Saito, H. (2005). Plantlet regeneration from protoplasts of *Muscari armeniacum* Leichtl. Ex Bak. *Plant Biotechnology*, 22(3), 249-251.

Namdeo, A.G. (2007). Plant cell elicitation for production of secondary metabolites: A review. *Pharmacognosy Reviews*, 1(1), 69-79.

Naquvi, K.J., Ali, M. and Ahamad, J. (2012). Antidiabetic activity of aqueous extract of *coriandrum sativum* L. fruits in streptozotocin induced rats. *International Journal of Pharmacy and Pharmaceutical Science*, 4(1), 239-240.

Navratilova, B. (2004). Protoplast cultures and protoplast fusion focused on Brassicaceae. *Horticulture Science*, 31, 140-157.

Navratilova, B., Rokytova, L. and Lebeda, A. (2000). Isolation of mesophyll protoplasts of *Cucumis* spp. and *Cucurbita* spp. *Acta Horticulturae*, 510, 425-431.

Neha, M.P. V., Suganthi, V. and Gowri, S. (2013). Evaluation of anti-inflammatory activity in ethanolic extract of *Coriandrum sativum* L. using carrageenan induced paw oedema in albino rats. *Der Pharma Chemica*, 5(2), 139-143.

Nwauzoma, A.B. and Jaja, E.T. (2013). A review of somaclonal variation in plantain (*Musa* spp): mechanisms and applications. *Journal of Applied Bioscience*, 67, 5252-5260.

Oksman-Caldentey, K.M. and Inze, D. (2004). Plant cell factories in the post-genomic era: new ways to produce designer secondary metabolites. *Trends in Plant Science*, 9, 433-440.

Ondrej, V., Kitner, M., Dolezalova, I., Nadvornık, P., Navratilova, B. and Lebeda, A. (2009). Chromatin structural rearrangement during dedifferentiation of protoplasts of *Cucumis sativus* L. *Moleculer Cells*, 27, 443-447.

Ou, B., Huang, D., Hampsch-Woodill, M., Flanagan, J.A. and Deemer, E.K. (2002). Analysis of antioxidant activities of common vegetables employing oxygen radical absorbance capacity (ORAC) and ferric reducing antioxidant power (FRAP) assays: A comparative study. *Journal of Agriculture Food Chemistry,* 50, 3122-3128.

Pal, S.P., Alam, I., Anisuzzaman, M., Sarker, K.K., Sharmin, S.A. and Alam, M.F. (2007). Indirect organogenesis in summer squash (*Cucurbita pepo* L.). *Turkish Journal of Agriculture and Forestry*, 31, 63-70.

Palanyandy, S.R., Gantait, S., Suranthran, P., Sinniah, U.R. and Subramaniam, S. (2015). Storage of encapsulated oil palm polyembryoids: influence of temperature and duration. *In Vitro Cellular and Developmental Biology - Plant,* 51, 118-124.

Pan, Z.G., Liu, C.Z., Zobayed, S.M. A. and Saxena, P.K. (2004). Plant regeneration from mesophyll protoplasts of *Echinacea Purpurea*. *Plant Cell, Tissue and Organ Culture*, 77(3), 251-255.

Patel, H. and Krishnamurthy, R. (2013). Elicitors in plant tissue culture. *Journal of Pharmacognosy and Phytochemistry*, 2(2), 60-65.

Pathak, N.L., Kasture, S.B., Bhatt, N.M. and Rathod, J.D. (2011). Phytopharmacological properties of *Coriandrum sativum* as a potential medicinal tree: An overview. *Journal of Applied Pharmaceutical Science*, 1(4), 20-25.

Pati, P.K., Sharma, M. and Ahuja, P.S. (2008). Protoplast isolation and culture. In: Keshavachandran, R. and Peter, K.V. (Eds.), *Plant biotechnology: methods in tissue culture and gene transfer* (pp. 115-132). India, Universities Press Pvt Ltd.

Paur, I., Carlsen, M.H., Halvorsen, B.L. and Blomhoff, R. (2011). Antioxidants in herbs and spices: Roles in oxidative stress and redox signalling. In: Benzie, I.F. F. and Wachtel-Galor, S. (Eds.), *Herbal Medicine: Biomolecular and Clinical Aspects*. Vol 2, Taylor and Francis, USA.

Pavlovic, S., Vinterhalter, B., Zdravkovic-Korac, S., Vinterhalter, D., Zdravkovic, J., Cvikic, D. and Mitic, N. (2012). Recurrent somatic embryogenesis and plant regeneration from immature zygotic embryos of cabbage (*Brassica oleracea* var. capitata) and cauliflower (*Brassica oleracea* var. botrytis). *Plant Cell, Tissue and Organ Culture*, 113, 497-506.

Potter, T.L. and Fagerson, I.S. (1990). Composition of coriander leaf volatiles. *J. Agric. Food Chem.* 38, 2054-2056.

Potters, G., Pasternak, T., Guisez, Y., Palme, K.J. and Jansen, M.A. K. (2007). Strees-induced morphogenic responses: growing out of trouble? *Trends in Plant Science*, 12(3), 98-105.

Poulev, A., O'Neal., J.M., Logendra, S., Pouleva, R.B., Timeva, V., Garvey, A.S., Gleba, D., Jenkins, I.S., Halpern, B.T., Kneer, R., Cragg, G.M. and Raskin, I. (2003). Elicitation, a new window into plant chemodiversity and phytochemical drug discovery. *Journal of Medicinal Chemistry*, 46(12), 2542-2547.

Prakash, B., Shukla, R., Singh, P., Kumar, A., Mishra, P.K. and Dubey, N.K. (2010). Efficacy of chemically characterized *Piper betle* L. essential oil against fungal and aflatoxin contamination of some edible commodities and its antioxidant activity. *International Journal of Food Microbiology*, 142, 114-119.

Prakash, B., Shukla, R., Singh, P., Mishra, P.K., Dubey, N.K. and Kharwar, R.N. (2011). Efficacy of chemically characterized *Ocimum gratissimum* L. essential oil as an antioxidant and a safe plant based antimicrobial against fungal and aflatoxin B1 contamination of spices. *Food Research International*, 44, 385-390.

Prakash, V. (1990). *Leafy Spices* (pp. 31-32). Boca Raton, CRC Press Inc.

Pramono, E. (2002). The commercial use of traditional knowledge and medicinal plants in Indonesia. *Multi-stakeholder dialogue on trade, intellectual property and biological resources in Asia* (pp. 1-13). Bangladesh, BRAC Rajendrapur.

Prange, A., Bartsch, M., Meiners, J., Serek, M. and Winkelmann, T. (2012). Interspecific somatic hybrids between *Cyclamen persicum* and *C. coum*, two sexually incompatible species. *Plant Cell Reports*, 31, 723-735.

Prange, A.N. S., Serek, M., Bartsch, M. and Winkelmann, T. (2010). Efficient and stable regeneration from protoplasts of *Cyclamen coum* Miller via somatic embryogenesis. *Plant Cell, Tissue Organ Culture*, 101(2), 171-182.

Pulianmackal, A.J., Kareem, A.V.K., Durgaprasad, K., Trivedi, Z.B. and Prasad, K. (2014). Competence and regulatory interactions during regeneration in plants. *Front. Plant Sci.* 5, doi: 10.3389/fpls.2014.00142

Radic, S., Prolic, M., Pavlica, M. and Pevalek-kozlina, B. (2005). Cytogenetic stability of *Centaurea ragusina* long-term culture. *Plant Cell, Tissue and Organ Culture*, 82, 343-348.

Radman, R., Saez, T., Bucke, C. and Keshavarz. T. (2003). Elicitation of plant and microbial systems. *Biotechnology and Applied Biochemistry*, 37, 91-102.

Rai, M.K., Asthana, P., Singh, S.K., Jaiswal, V.S. and Jaiswal, U. (2009). The encapsulation technology in fruit plants- a review. *Biotechnology Advances*, 27, 671-679.

Raikar, S.V., Braun, R.H., Bryant, C., Conner, A.J. and Christey, M.C. (2008). Efficient isolation, culture and regeneration of *Lotus corniculatus* protoplast. *Plant Biotechnology Reports*, 2(3), 171-177.

Rakoczy-Trojanowska, M. (2002). Alternative methods of plant transformation- a short review. *Cellular and Molecular Biology Letters*, 7, 849-858.

Ramadan, M.F. and Morsel, J.T. (2003). Analysis of glycolipids from black cumin (*Nigella sativa* L.), coriander (*Coriandrum sativum* L.) and niger (*Guizotia abyssinica* Cass.) oil seeds. *Food Chemistry*, 80, 197-204.

Ramakrishna, A. and Ravishankar, G.A. (2011). Influence of abiotic stress signals on secondary metabolites in plants. *Plant Signaling and Behaviour*, 6(11), 1720-1731.

Rao, M.R., Palada, M.C. and Becker, B.N. (2004). Medicinal and aromatic plants in agro-forestry systems. *Agroforestry Systems*, 6, 1107-1122.

Rao, N.K. (2004). Plant genetic resources: Advancing conservation and use through biotechnology. *African Journal of Biotechnology*, 3, 136-145.

Rao, R.S. and Ravishankar, G.A. (2002). Plant cell cultures: chemical factories of secondary metabolites. *Biotechnology Advances*, 20, 101-153.

Rastogi, S.C. (2003). *Cell and Molecular Biology*. United States of America: New Age International.

Raut, J.S. and Karuppayil, S.M. (2014). A status review on the medicinal properties of essential oils. *Industrial Crops and Products*, 62, 250-264.

Rewers, M., Drouin, J., Kisiala, A., Sliwinska, E. and Cholewa, E. (2012). In vitro regenerated wetland sedge *Eriophorum vaginatum* L. is genetically stable. *Acta Physiology Plantarum*, 34, 2197-2206.

Rezazadeh, R. and Niedz, R.P. (2015). Protoplast isolation and plant regeneration of guava (*Psidium guajava* L.) using experiments in mixture-amount design. *Plant Cell, Tissue and Organ Culture*, 122, 585-604.

Ribkahwati, Purnobasuki, H., Isnaeni and Utami, E.S. W. (2015). Quantity essential oil from rose callus leaf (*Rosa hybrid* L. variety *Hybride tea purple*): Results of light elicitation. *Journal of Chemical and Pharmaceutical Research*, 7(4), 496-499.
Roat, C. and Ramawat, K.G. (2009). Elicitor induced accumulation of stilbenes in cell suspension cultures of *Cayratia trifoliata* (L.) Domin. *Plant Biotechnology Reports*, 3, 135-138.
Robert, D.R., Flinn, B.S., Webb, D.T., Webster, F.B. and Sutton, B.C. S. (1986). Characterization of immature embryos of interior spruce by SDS-PAGE and microscopy in relation to their competence for somatic embryogenesis. *Plant Cell Reports*, 8, 285-288.
Santos, M.O., Romano, E., Yotoko, K.S.C., Tinoco, M.L.P., Dias, B.B. A. and Aragao, F.J. L. (2005). Characterization of the cacao somatic embryogenesis receptor-like kinase (SERK) gene expressed during somatic embryogeesis. *Plant Science*, 168, 723-729.
Schuler, P., 1990. Natural antioxidants exploited commercially. In: Hudson, B.J.F. (Ed.), *Food Antioxidants* (pp. 99-170). London, Elsevier.
See, K.S., Bhatt, A. and Keng, C.L. (2011). Effect of sucrose and methyl jasmonate on biomass and anthocyanin production in cell suspension culture of *Melastoma malabathricum* (Melastomaceae). *Revista de Biologia Tropical*, 59(2), 597-606.
Shah, A.H., Rasheed, N., Haider, M.S., Saleem, F., Tahir, M. and Iqbal, M. (2009). An efficient, short and cost effective regeneration system for transformation studies of sugarcane (*Saccharum officinarum* L.). *Pakistan Journal of Botany*, 41, 609-614.
Sharma, A., Rathour, R. and Plaha, P. (2010). Induction of *Fusarium* wilt (*Fusarium oxysporum* f. sp. pisi) resistance in garden pea using induced mutagenesis and in vitro selection techniques. *Euphytica*, 173, 345-356.
Sharmin, S.A., Alam, M.J., Sheikh, M.M.I., Sarker, K.K., Khalekuzzaman, M., Haque, M.A., Alam, M.F. and Alam, F. (2014). Somatic embryogenesis and plant regeneration in *Wedelia calendulacea* Less. an endangered medicinal plant. *Brazilian Archives of Biology and Technology*, 57(3), 394-401.

Shimizu, Y., Maeda, K., Kato, M. and Shimomura, K. (2010). Methyl jasmonate induces anthocyanin accumulation in *Gynura bicolor* cultured roots. *In Vitro Cellular and Developmental Biology - Plant*, 46, 460-465.

Silva, F., Ferreira, S., Duarte, A., Mendonca, D.I. and Domingues, F.C. (2011). Antifungal activity of *Coriandrum sativum* L. essential oil, its mode of action against *Candida* species and potential synergism with amphotericin B. *Phytomedicine*, 19, 42-47.

Sliwinska, E. and Thiem, B. (2007). Genome size stability in six medicinal plant species propagated in vitro. *Biologia Plantarum*, 51, 556-558.

Small, E. (1997). *Culinary Herbs* (pp. 219-225). Ottawa, NRC Research Press.

Smallfield, B.M., Van-Klink, J.W., Perry, N.B. and Dodds, G. (2001). Coriander spice oil: effects of fruit crushing and distillation time on yield and composition. *Journal of Agricultural and Food Chemistry*, 49, 118-123.

Smetanska, I. (2008). Production of secondary metabolites using plant cell cultures. *Advances in Biochemical Engineering Biotechnology*, 111, 187-228.

Snyman, S.J., Meyer, G.M., Koch, A.C., Banasiak, M. and Watt, M.P. (2011). Applications of in vitro culture systems for commercial sugarcane production and improvement. *In Vitro Cellular and Developmental Biology - Plant*, 47, 234-249.

Strauss, S.H., Rottman, W.H., Brunner, A.M. and Sheppard, L.A. (1995). Genetic emgineering of reproductive sterility in forest trees. *Molecular Breeding*, 1, 5-26.

Swamy, N.R., Ugandhar, T., Praveen, M., Venkataiah, P., Rambabu, M., Upender, M. and Subhash, K. (2005). Somatic embryogenesis and plantlet regeneration from cotyledon and leaf explants of *Solanum surattense*. *Indian Journal of Biotechnology*, 4, 414-418.

Sylvestre, M., Legault, J., Dufour, D. and Pichette, A. (2005). Chemical composition and anticancer activity of leaf essential oil of *Myrica gale* L. *Phytomedicine*, 12, 299-304.

Sylvestre, M., Pichette, A., Longtin, A., Nagau, F. and Legault, J. (2006). Essential oil analysis and anticancer activity of leaf essential oil of *Croton flavens* L. from Guadeloupe. *Journal of Ethnopharmacology*, 103, 99-102.

Szopa, A., Ekiert, H., Szewczyk, A. and Fugas, E. (2012). Production of bioactive phenolic acids and furanocoumarins in in vitro cultures of *Ruta graveolens* L. and *Ruta graveolens* ssp. *Divaricata* (Tenore) Gams under different light conditions. *Plant Cell, Tissue and Organ Culture*, 110, 329-336.

Tahami, S.K., Chamani, E. and Zare, N. (2014). Plant regeneration from protoplasts of *Lilium ledebourii* (Baker) Boiss. *Journal of Agricultural Science and Technology*, 16, 1133-1144.

Taiz, L. and Zeiger, E. (2004). *Fisiologia Vegetal* (pp 719), third ed. Artmed, Porto Alegre.

Tamer, M.I. and Mavituna, F. (1996). Protease from callus and cell suspension cultures of *Onopordum turcicum* (Compositae). *Biotehcnology Letters*, 18, 361-366.

Taniguchi, M., Yanai, M., Xiao, Y.Q., Kido, T. and Baba, K. (1996). Three isocoumarins from *Coriandrum sativum*. *Phytochemistry*, 42(3), 843-846.

Teixeira, B., Marques, A., Ramos, C., Neng, N.R., Nogueira, J.M., Saraiva, J.A. and Nunes, M.L. (2013). Chemical composition and antibacterial and antioxidant properties of commercial essential oils. *Industrial Crops and Products*, 43, 587-595.

Thanh, N.T., Murthy, H.N., Yu, K.W., Hahn, E.J. and Peak, K.Y. (2005). Methyl jasmonate elicitation enhanced synthesis of ginsenoside by cell suspension culture of Panax ginseng in 5L balloon type bubble bioreactor. *Applied Microbiology and Biotechnology*, 67(2): 197-201.

Thiem, B., Kikowska, M., Krawczyk, A., Wiekowska, B. and Sliwinska, E. (2013). Phenolic acid and DNA contents of micropropagated *Eryngium planum* L. *Plant Cell, Tissue and Organ Culture*, 114, 197-206.

Thorpe, T.A. (2007). History of plant tissue culture. *Molecular Biotechnology*, 37, 169-1 80.

Tilman, D., Cassman, K.G., Matson, P.A., Naylor, R. and Polasky, S. (2002). Agricultural sustainability and intensive production practices. *Nature*, 418, 671-677.

Tiwari, J.K., Sarkar, D., Pandey, S.K., Gopal, J. and Kumar, S.R. (2010). Molecular and morphological characterization of somatic hybrids between *Solanum tuberosum* L. and *S. Etuberosum* Lindl. *Plant Cell, Tissue and Organ Culture*, 103(2), 175-187.

Tomiczak, K., Mikuła, A., Sliwinska, E. and Rybczyński, J.J. (2015). Autotetraploid plant regeneration by indirect somatic embryogenesis from leaf mesophyll protoplasts of diploid *Gentiana decumbens* L.f.*In Vitro Cellular and Developmental Biology - Plant*, 51(3), 350-359.

Tudses, N., Premjet, S. and Premjet, D. (2014). Optimal conditions for high-yield protoplast isolations of *Jatropha curcas* L. and *Ricinus communis* L. *American-Eurasian Journal of Agricultural and Environmental Science*, 14(3), 221-230.

Valladares, S., Sanchez, C., Martinez, M.T., Ballester, A. and Vieitez, A.M. (2006). Plant regeneration through somatic embryogenesis from tissues of mature oak trees; true to type conformity of plantlets by RAPD analysis. *Plant Cell Reports*, 25, 879-886.

Verdeil, J.L., Alemanno, L., Niemenak, N. and Trangarder, T.J. (2007). Pluripotent versus totipotent plant stem cells: dependence versus autonomy. *Trends in Plant Science*, 12, 245-252.

Vigan, M. (2010). Essential oils: renewal of interest and toxicity. *Eurasian Journal of Dermatology*, 20, 685-692.

Voon, C.H., Bhat, R. and Rusul, G. (2012). Flower extracts and their essential oils as potential antimicrobial agents for food uses and pharmaceutical applications. *Comprehensive Reviews in Food Science and Food Safety*, 11, 34-55.

Wasternack, C. and Hause, B. (2002). Jasmonates and octadecanoids: signals in plant stress responses and development. *Progress in Nucleic Acid Research and Molecular Biology*, 72, 165-221.

Wiszniewska, A. and Pindel, A. (2009). Improvement in *Lupinus luteus* (Fabaceae) protoplast cultures- stimulatory effect of agarose embedding

and chemical nursing on protoplast divisions. *Australian Journal of Botany*, 57, 502-511.
Wongwicha, W., Tanaka, H., Shoyama, Y. and Putalun, W. (2011). Methyl jasmonate elicitation enhances glycyrrhizin production in *Glycyrrhiza inflata* hairy roots cultures. *Zeitschrift fur Naturforschung C*, 66, 423-428.
Yentema, O., Alioune, O. and Dorosso, S.A. (2007). Chemical composition and physical characteristics of the essential oil of *Cymbopogon schoenanthus* (L.) sppreng of Burkina faso. *Journal of Applied Sciences*, 7(4), 503-507.
Yeong, H.Y., Khalid, N. and Phang, S.M. (2008). Protoplast isolation and regeneration from *Gracilaria changii* (Gracilariales, Rhodophyta). *Journal of Applied Phycology,* 20, 641-651.
Yoo, S.D., Cho, Y.H. and Sheen, J. (2007). Arabidopsis mesophyll protoplasts: A versatile cell system for transient gene expression analysis. *Nature Protocol*, 2(7), 1565-1572.
Zabala, M.A., Angarita, M., Restrepo, J.M., Caicedo, L.A. and Perea, M. (2010). Elicitation with methyl-jasmonate stimulates peruvoside production in cell suspension cultures of *Thevetia peruviana*. *In Vitro Cellular and Developmental Biology - Plant*, 46, 233-238.
Zavattieri, M.A., Frederico, A.M., Lima, M., Sabino, R. and Arnholdt-Schmitt, B. (2010). Induction of somatic embryogenesis as an example of stress-related plant reactions. *Electronic Journal of Biotechnology*, 13(1), doi: 10.2225/vol13-issue1-fulltext-4.
Zhai, Z., Sooksa-nguan, T. and Vatamaniuk, O.K. (2009). Establishing RNA interference as a reverse-genetic approach for gene functional analysis in protoplasts. *Plant Physiology*, 149, 642-652.
Zhang, J., Shen, W., Yan, P., Li, X. and Zhou, P. (2011). Factors that influence the yield and viability of protoplasts from *Carica papaya* L. *African Journal of Biotechnology*, 10(26), 5137-5142.
Zhao, J., Davis, L.C. and Verpoorte, R. (2005). Elicitor signal transduction leading to production of plant secondary metabolites. *Biotechnology Advances*, 23, 283-294.

Zheng, S.J., Henken, B., De Maagd, R., Purwito, A., Krens, F. and Kik, C. (2005). Two different *Bacillus thuringiensis* toxin genes confer resistance to beet armyworm (Spodoptera exigua Hubner) in transgenic Bt-shallots (*Allium cepa* L.). *Transgenic Research*, 14, 261-272.

Zimmerman, J.L. (1993). Somatic Embryogenesis: A model for early development in higher plants. *Plant Cell*, 5(10), 1411-1423.

Zu, Y.G., Yu, H.M., Liang, L., Fu, Y.J., Efferth, T., Liu, X. and Wu, N. (2010). Activities of ten essential oils towards *Propionibacterium acnes* and PC-3, A-549 and MCF-7 cancer cells. *Molecules*, 15, 3200-3210.

In: Coriander
Editor: Deepak Kumar Semwal

ISBN: 978-1-53616-483-1
© 2019 Nova Science Publishers, Inc.

Chapter 5

CORIANDER SEED AS A SOURCE OF BIOLOGICALLY ACTIVE COMPOUNDS: ESSENTIAL OIL, CHEMICAL COMPOSITION AND BIOLOGICAL ACTIVITY

Saša Đurović* and Stevan Blagojević
Institute of General and Physical Chemistry,
Belagrde, Republic of Serbia

ABSTRACT

Coriander (*Coriandrum sativum* L.) is an annual herb from the *Apiaceae* botanical family. It is widely distributed and cultivated in Mediterranean countries for the seeds. The plant is usually used as food and in pharmaceutical and cosmetic industries. Many different studies showed that coriander and its extracts express various activities such as antioxidant, cytotoxic, antimicrobial, anticarcinogenic, antimutagenic and antidiabetic activity. Such a wide range of activity seeds possesses by virtue of its chemical composition. Linalool, geraniol, camphor and limonene are the main compounds in essential oil. Aldehyde and alcohols,

*Corresponding Author's Email: sasatfns@uns.ac.rs.

fatty acids, sterols and polyphenolic compounds have also been found. In order to achieve a maximal result of applied processes, two approaches have been applied (response surface methodology and artificial neural network) on different extraction techniques. Summarizing, results have shown that this plant justifiably possesses one of the most significant places in pharmaceutical and food industries in these days.

Keywords: coriander seeds, essential oil, extraction, chemical composition, optimization, biological activity

INTRODUCTION

Coriander (*Coriandrum sativum* L.) is an annual herb from the *Apiaceae* botanical family. It is widely distributed and cultivated in Mediterranean countries for the seeds (Eikani, Golmohammad, & Rowshanzamir, 2007; Grosso et al., 2008; Pavlić, Vidović, Vladić, Radosavljević, & Zeković, 2015; Wangensteen, Samuelsen, & Malterud, 2004), which are spherically shaped, longitudinally striated and slightly pointed at one end. Length is 3-5 mm, usually brown coloured (when dried), but it may be green, straw-coloured or off white (Coşkuner & Karababa, 2007). They are commercially usually available in ground powder form or in whole. It possesses sweet aromatic smell with a slightly spicy bittersweet taste and is traditionally used in cooking in regions such as South Asia, North Africa, Latin America and Europe (Dima, Ifrim, Coman, Alexe, & Dima, 2016).

Plants synthesize a broad range of primary and secondary metabolites during their life cycle. These compounds proved to be biologically active and thus attract the scientific community for a while. Extracts of plants containing these compounds are often used in food, pharmaceutical and cosmetic industries. In order to obtain extract different extraction techniques have been developed (Azmir et al., 2013; Wang & Weller, 2006).

In the beginning, conventional techniques have appeared. In this group simple liquid-solid, liquid-liquid, maceration hydrodistillation and Soxhlet extraction techniques are included. These techniques have been used for a long time and included the application of higher temperature and large amounts of organic solvents which are usually toxic. It is also worthy of mention that these techniques are time-consuming i.e., requires several hours or days in some cases to complete extraction or just to achieve satisfactory yield (Azmir et al., 2013; da Silva, Rocha-Santos, & Duarte, 2016; Radojković et al., 2016; Wang & Weller, 2006). For such reasons, nonconventional techniques have been developed. Maceration and soxhlet extraction are successfully changed with ultrasound-assisted, microwave-assisted and subcritical water extraction techniques, while hydrodistillation has been changed with supercritical fluid extraction for isolation of moderately and nonpolar compounds from plant material (Cvetanović et al., 2017; P. Mašković et al., 2018, 2017; P. Z. Mašković et al., 2018; Veličković et al., 2017).

Aim of this chapter is to give insight into the application of extraction techniques for isolation of biomolecules from the coriander seed. Chemical composition of obtained extracts will be given as well as the optimization of the extraction process using response surface methodology (RSM) and artificial neural network (ANN).

EXTRACTION TECHNIQUES

Extraction techniques may be divided into two major groups :

1. Conventional extraction techniques (maceration, soxhlet extraction, hydrodistillation, etc.)
2. Nonconventional extraction techniques (ultrasound-assisted, microwave-assisted, supercritical fluid and subcritical water extraction)

Conventional Extraction Techniques

All conventional techniques are based on the application of usually organic solvent in combination with heat and/or agitation (Wang & Weller,

2006). Hydrodistillation is usually used for isolation of volatile compounds from plants (terpenoids for example). Using a Clevenger apparatus (Figure 1, left), volatile compounds are isolated by water vapour. Round flask is filled with plant material and water and heated. Created vapour carries volatile compounds, which are further condensed in a condenser and collected in an organic solvent (usually *n*-hexane). Condensed water returns into the flask through pipe vaporizing again (Đurović, 2019).

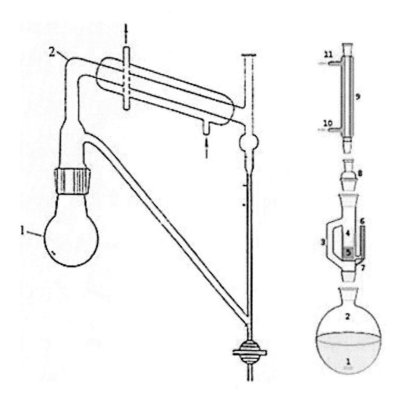

Figure 1. Apparatus for isolation of volatile compounds and for Soxhlet extraction. Clevenger apparatus (left): (1) round flask, (2) Clevenger apparatus with condenser. Soxhlet extraction (right): (1) solvent, (2) round flask, (3) side tube, (4) extractor, (5) shell for plant material, (6) syphon, (7) recirculation tube, (8) extension, (9) condenser, (10) coolant in, (11) coolant out (Đurović, 2019).

The soxhlet extraction technique is used for a long time and represents a reference technique for evaluation of the performance of other extraction techniques (Wang & Weller, 2006; Zeković, Cvetanović, et al., 2017).

This technique usually surpasses other extraction in performance but is limited by the thermal stability of natural compounds. Apparatus for Soxhlet extraction is given in Figure 1 (right). It is comprised of a round flask filled with solvent, Soxhlet extractor with a shell filled with plant material and condenser. When condensed solvent filled thimble-solder till overflow level, it is returning to the flask through the syphon. This represents on cycle and process may be repeated in several cycles until the extraction completion (Wang & Weller, 2006).

Maceration is used for preparations of tonics in homes for a long time. It is comprising of several steps: grinding plant material, the immersion of ground material into the desired solvent, removing the solvent and pressing the mark to recover occluded solution. Maceration is taking place in closed vessels with occasional mixing for two reasons (Azmir et al., 2013):

- The increasing diffusion of desired compounds from plant material to solvent,
- Removing the concentrated solution from the surface of plant material which actually increases extraction yield.

Although conventional techniques are still widely applied, they have many disadvantages. Among them, the application of toxic and environmental non-friendly solvents, long duration of extraction process and the possibility of degradation of desired compounds. To overcome these problems, nonconventional extraction techniques are introduced to the scientific community.

Nonconventional Extraction Techniques

Main nonconventional extraction techniques are ultrasound-assisted, microwave-assisted, subcritical water and supercritical fluid extraction. Every technique relies on different effects which are responsible for the technique's selectivity and/or effectiveness.

Ultrasound-assisted extraction (UAE) uses ultrasound wave, which is beyond human hearing. Applied frequency in chemistry is usually 20 kHz to 100 MHz. Ultrasound passes through medium provoking compression and expansion. This phenomenon is known as cavitation. Cavitation includes production, growth and explosion of bubbles, where a large amount of energy (Azmir et al., 2013; Farid Chemat, Tomao, & Virot, 2008; Rostagno, Palma, & Barroso, 2003; Zekovic, Djurovic, & Pavlic, 2016). Beside cavitation, there are also mechanical effects which enhance penetration of solvent into cellular material (Rostagno et al., 2003; Zekovic et al., 2016). This technique is cheap, offering high economic efficiency, low equipment requirements, high reproducibility, simplified manipulation and low solvent and energy consumption (Zekovic et al., 2016). Disadvantages of this process are induced changes in the chemical composition and occurrence of degradation of desired compounds throughout the formation of free radical species inside the gas bubbles (Paniwnyk, Beaufoy, Lorimer, & Mason, 2001; Zekovic et al., 2016).

Microwave-assisted extraction (MAE) with a frequency of 300 MHz to 300 GHz offers rapid delivery of energy, thus matrix is heating more efficiently and homogeneously (Wang & Weller, 2006). Heating occurs due to ion conduction and dipole rotation mechanisms, where during the ionic conduction mechanism resistance of the medium to ion flow occurs and generates heat (Azmir et al., 2013; Jain, Jain, Pandey, Vyas, & Shukla, 2009; Zeković, Cvetanović, et al., 2017). The extraction process involves three sequential steps as follows (Azmir et al., 2013):

- Separation of solutes from active sites under increased pressure and temperature,
- Diffusion of solvent across sample matrix,
- Release of solutes from the sample matrix to the solvent.

MAE offers several advantages comparing to other extraction techniques: quicker heating, reduced thermal gradients, the smaller size of the equipment and increased extraction yield (Cravotto et al., 2008). Besides these, one more very significant advantage is the possibility of usage of green solvents, such as water, for the extraction process.

Subcritical water extraction (SWE) represents modern extraction technique which attracts more and more attention of the scientific community. This technique relies on changing dielectric constant (ε) of water with temperature. At ambient conditions, water is polar solvent with ε ≈ 80, but with an increase in temperature, ε decreases. For example, at 110°C dielectric constant of water is 53 and at 190°C is 36.5, which is a similar value to that of methanol ($\varepsilon = 32.6$ at 25°C). To keep water in a liquid state, high pressure is applied (≈10 MPa) (Cvetanović et al., 2016; Ko, Cheigh, & Chung, 2014). Taking into account such behaviour of subcritical water, by increasing the operation temperature selectivity of the extraction process moves toward moderately polar compounds and then non-polar. Previously conducted studies showed that both temperature and pressure have an impact on extraction efficiency and selectivity, but temperature expressed stronger impact than pressure (Cvetanović et al., 2018, 2019). Typical equipment for SWE contains following parts: nitrogen cylinder, manometer, input gas valve, thermometer, extraction vessel able to hold higher pressures, a vibrating platform for shaking and heating plate as previously described (Cvetanović et al., 2016).

Supercritical fluid extraction (SFE) is green extraction method introduced as an alternative technique to hydrodistillation and solid-liquid extraction for isolation of essential oil from plant material (da Silva et al., 2016; Đurović, Šorgić, Popov, Radojković, & Zeković, 2018; Zeković, Bera, Đurović, & Pavlić, 2017). Development of SFE technique has started in 1980s, while industrial application started with decaffeination of green coffee beans and black tea leaves. It continues with the isolation of essential oils, oleoresins, and flavouring compounds from natural sources. It is also used for fractionation of edible oils and removal of pesticides from plant material (Brunner, 1994, 2005; Đurović et al., 2018; Jesus & Meireles, 2014). This technique is based on the introduction of fluids into a supercritical state, which is achieved by tuning temperature and pressure above critical values for a given fluid. In this state, fluid possesses properties between gaseous and liquid state. This means that the density of the supercritical fluid is similar to the density of liquids, but viscosity is similar to the gaseous state. Such properties influence on the diffusivity of supercritical fluid, which ranges between gaseous and liquid state furtherly resulted in better transport properties (Herrero, Cifuentes, & Ibanez, 2006; Meullemiestre, Breil, Abert-Vian, & Chemat, 2015). Properties of supercritical fluids can be easily modulated by tuning pressure and temperature of the system. This directly influences the density of the fluid and indicating changes in solubility of desired compounds from plant material (de Melo, Silvestre, & Silva, 2014).

The most common fluid for SFE is carbon-dioxide. There are several reasons for its selection. For instance, it is cheap, nontoxic, nonexplosive, it is not inflammable, chemically is inactive and possesses moderate critical parameters (p_c = 73.8 bar and t_c = 31.1°C) (de Melo et al., 2014; Đurović et al., 2018; Radojković et al., 2016; Zeković, Bera, et al., 2017). Typical equipment for SFE (Figure 2) contains the gas cylinder, compressor, extractor, separator, heat exchanger, thermostat, valves for flow regulation.

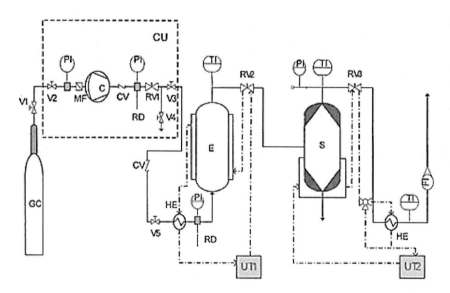

Figure 2. Laboratory-scale SFE plant. (GC) gas cylinder, (CU) compressor unit, (C) compressor with diaphragm, (E) extractor, (S) separator, (HE) heat exchanger, (UT) ultra-thermostat, (RV) regulation valve, (V) on/off valve, (MF) microfilter, (CV) cut-off valve, (RD) rupture disc, (PI) pressure indicator, (TI) temperature indicator, (FI) flow indicator (Filip et al., 2016).

EXTRACTION OF CORIANDER SEED AND OPTIMIZATION OF THE PROCESS

Literature indicated that both conventional and nonconventional extraction techniques applied for isolation of bioactive compounds from coriander (Grosso et al., 2008; Pavlić et al., 2015; Zeković, Bera, et al., 2017; Zeković et al., 2015, 2014; Zekovic et al., 2016; Zeković, Pavlić, Cvetanović, & Đurović, 2016; Zeković, Vladić, Vidović, Adamović, & Pavlić, 2016). Total extraction yields are shown in Table 1.

Pavlic et al., (Pavlić et al., 2015) and Zekovic et al., (Zeković et al., 2015) showed that Soxhlet extraction achieved the highest total extraction yield. It has been showed also that particle size (d) has a significant impact on extraction process (Zeković et al., 2015). Presented results indicated that smaller particles have been better extracted (total extraction yield is higher). Pavlić et al., (Pavlić et al., 2015) indicated that Soxhlet extraction achieved a higher yield than subcritical water extraction (0.36-2.22%), but it is also demonstrated that SWE technique gives better result than hydrodistillation when it comes to the isolation of essential oils (Eikani, Golmohammad, & Rowshanzamir, 2007).

Table 1. Total extraction yield of hydrodistillation and Soxhlet extraction of coriander seeds

Technique	Conditions	Yield (%, w/w)
Hydrodistillation*	t < 100°C, 2 h	0.600
Hydrodistillation**	t < 100°C, 2 h, d_1	0.350
	t < 100°C, 2 h, d_2	0.575
	t < 100°C, 2 h, d_3	0.600
Soxhlet extraction*	Boiling point, 15 exhanges	14.450
Soxhlet extraction**	Boiling point, 15 exhanges, d_1	2.540
	Boiling point, 15 exhanges, d_2	8.100
	Boiling point, 15 exhanges, d_3	16.770

*Result from Pavlic et al., (Pavlić et al., 2015), **Results published by Zekovic et al., (Zeković et al., 2015), d_1 = 1.368 mm, d_2 = 0.775 mm, d_3 = 0.631 mm.

Beside conventional extraction techniques, nonconventional techniques have been also applied. UAE and MAE extractions have been used for isolation of moderately polar compounds such as phenolic and polyphenolic compounds (Gallo, Ferracane, Graziani, Ritieni, & Fogliano, 2010; Zekovic et al., 2016; Zeković, Vladić, et al., 2016). Gallo et ail. (Gallo et al., 2010) have analyzed obtained extracts and reported phenolic content of 41.812 and 82.091 mg GA/100 g (GA-gallic acid equivalent) for UAE and MAE, respectively. Zeković et al., optimized both UAE (Zekovic et al., 2016) and MAE (Zeković, Vladić, et al., 2016) extraction techniques for isolation of polyphenolic compounds from coriander. For the optimization process response surface methodology (RSM) has been applied. Optimization of any process is quite an important step, which aims to gain a maximum of the process itself but with minimal losses at the same moment. In this case, the subject of optimization is operation conditions for those two extraction techniques. In the case of UAE, ultrasonic power, time and temperature were optimized (Zekovic et al., 2016), while irradiation power, ethanol concentration and extraction time were optimized in the case of MAE (Zeković, Vladić, et al., 2016). RSM represents the most frequent allied technique for optimization. It is a collection of statistical and mathematical methods and is based on the influence of several variables on the response of interest with the final goal to optimize desired response (Adamczyk, Horny, Tricoteaux, Jouan, & Zadam, 2008; Baş & Boyacı, 2007). RSM also generates a mathematical model which may be able to successfully describe chemical processes within the given experimental conditions (Myers, Montgomery, & Anderson-Cook, 2009).

The most commonly used model in RSM is Box-Behnken model (BBD), which comprises of several numerical factors, three levels and minimum 15 randomized experiments with minimum 3 replicates in central point. Extraction parameters are independent variables, which are forced to range from -1 to +1. This is known as normalization and provides more evenly influence on response and also makes units of parameters to be irrelevant (Zeković, Vladić, et al., 2016).

In Table 2 are given natural and coded values of independent variables used in BBD in MAE and UAE extraction. From the presented parameters it might be noticed that MAE takes shorter than UAE, but irradiation power in MAE is quite higher as compared to ultrasonic power in the UAE process.

The third independent parameter was temperature (in the case of UAE) and ethanol concentration (in the case of MAE). Before optimization of the process, preliminary extractions have to be conducted in order to investigate the influence of parameters on the process. Another approach is to consult the available literature to gather available data. In both cases, available data helps to select which parameter will be optimized and in which range. Tables 3 and 4 show the results of optimization of UAE and MAE extractions.

In both cases, measured responses were total phenolics content (TPC), total flavonoids content (TFC), DPPH scavenging activity (IC_{50}) and reducing power (EC_{50}). Results showed that TPC varied from 221.50 to 374.25 mg GAE/100 g and 136.92 to 384.54 mg GAE/100g for UAE and MAE, respectively.

Ranges of TFC were 64.50-153.74 mg CE/100 g and 94.50-211.83 mg CE/100 g for UAE and MAE, respectively. MAE showed better results in the isolation of phenolic compounds and flavonoids from coriander seeds. Same trends are noticeable in the case of IC_{50} and EC_{50} values. IC_{50} ranges from 35.69 to 53.98 µg/mL for UAE and from 30.20 to 66.50 µg/mL for MAE. EC_{50} values are in both cases 10-fold higher than IC_{50} and ranges from 143.60 to 165.10 µg/mL for UAE and from 115.30 to 182.40 µg/mL for MAE. In these two responses, lower values indicate higher antioxidant activity. Taking this into account, MAE showed better results. This is in consistency with the previous studies which showed that concentration of TPC and TFC are in correlation with the antioxidant and cytotoxic activities of extracts (P. Mašković et al., 2017; P. Z. Mašković et al., 2018; Veličković et al., 2017).

Table 2. Natural and coded levels of independent variables in RSM

Extraction technique	Variable	Coded levels		
		-1	0	1
		Natural levels		
UAE*	Temperature (°C)	40	60	80
	Extraction time (min)	40	60	80
	Ultrasonic power	96	156	216
MAE**	Extraction time (min)	15	25	35
	Ethanol concentration (%)	50	70	90
	Irradiation power (W)	400	600	800

*Parameters from Zeković et al., (Zekovic et al., 2016), **Parameters from Zeković et al., (Zeković, Vladić, et al., 2016).

Table 3. Experimental conditions for the BBD design of UAE of coriander

Independent variable			Measured response*			
X_1	X_2	X_3	TPC (mg GAE/100 g)	TFC (mg CE/100 g)	IC_{50} (μg/mL)	EC_{50} (μg/mL)
0	0	0	287.76	150.40	43.44	152.30
0	-1	1	307.28	203.40	45.16	158.20
1	-1	0	350.82	203.90	53.98	165.10
0	0	0	282.67	160.40	47.98	148.30
0	1	1	296.61	162.90	51.05	157.70
-1	1	0	222.32	124.60	53.92	163.40
1	0	1	364.74	192.60	35.69	144.80
-1	-1	0	221.53	126.90	50.94	154.50
0	0	0	288.17	164.70	49.90	156.60
-1	0	1	260.05	145.70	48.20	155.20
0	0	0	265.60	155.40	48.89	158.40
0	-1	-1	326.05	190.50	52.12	165.00
1	0	-1	374.25	198.70	48.62	147.40
0	1	-1	310.64	165.01	52.52	158.30
-1	0	-1	240.85	138.90	48.81	148.30
1	1	0	372.10	199.50	48.68	143.60
0	0	0	287.46	149.30	48.86	14.64

*Results from Zeković et al., (2016).

Table 4. Experimental conditions for the BBD design of MAE of coriander

Independent variable			Measured response*			
X_1	X_2	X_3	TPC (mg GAE/100 g)	TFC (mg CE/100 g)	IC_{50} (µg/mL)	EC_{50} (µg/mL)
0	0	0	294.13	213.62	35.40	135.00
0	-1	1	346.35	201.79	66.50	153.20
1	-1	0	384.54	210.39	54.10	175.30
0	0	0	287.96	210.93	35.30	117.90
0	1	1	145.90	104.31	60.30	182.40
-1	1	0	136.92	94.50	49.90	169.00
1	0	1	291.33	211.83	46.00	130.20
-1	-1	0	358.15	201.26	38.10	161.70
0	0	0	284.03	198.57	35.20	115.30
-1	0	1	268.31	189.08	35.00	133.60
0	0	0	284.03	208.06	38.50	126.50
0	-1	-1	372.24	204.66	41.80	178.40
1	0	-1	299.19	214.69	36.40	1476.30
0	1	-1	154.42	109.91	55.30	157.70
-1	0	-1	250.90	202.15	30.20	143.70
1	1	0	163.31	120.45	53.90	159.50
0	0	0	279.54	202.69	35.60	127.40

*Results from Zeković et al., (Zeković, Vladić, et al., 2016).

Obtained results from the experiments are fitted into the second-order polynomial model. This model is able to describe the relationship between the investigated responses and independent variables:

$$Y = \beta_0 + \sum_{i=1}^{3}\beta_i X_i + \sum_{i=1}^{3}\beta_{ii}X_i^2 + \sum\sum_{i<j=1}^{3}\beta_{ij}X_iX_j \quad (1)$$

where Y is a measured response, β_0 is constant, b_j, b_{jj}, b_{ij} are the linear, quadratic and interactive coefficients of the model, respectively; X_i and X_j are the levels of the independent variables.

Applied software is Design-Expert v.7 Trial (State-Eset, Minneapolis, Minnesota, USA). Applied level of significance is 0.05, while adequacy of models has been evaluated by the coefficient of multiple determination (R^2), coefficient of variance (CV), p-values and lack of fit. These parameters are obtained by performing an analysis of variance (ANOVA) test in the mentioned software (Zekovic et al., 2016; Zeković, Vladić, et al., 2016). Examples of such report are given in Figures 3 and 4.

A report in Figure 3 showed that the applied model for TPC is significant ($p < 0.05$), while the lack of fit is not significant ($p > 0.05$). Analyzing the influence of other parameters, it might be noticed that A and C^2 showed significant influence or temperature and quadratic term of ultrasonic power. Coefficient of multiple determination is higher than 0.90 ($R^2 = 0.9709$). This indicates that the model represents a good approximation of the experimental results in the case of TPC.

Response 2 TP
ANOVA for Response Surface Quadratic Model
Analysis of variance table [Partial sum of squares - Type III]

Source	Sum of Squares	df	Mean Square	F Value	p-value Prob > F	
Model	36378.25	9	4042.03	25.94	0.0001	significant
A-Temperature	33431.81	1	33431.81	214.55	< 0.0001	
B-Time	2.01	1	2.01	0.013	0.9128	
C-Ultrasonic power	66.76	1	66.76	0.43	0.5337	
AB	104.96	1	104.96	0.67	0.4389	
AC	206.07	1	206.07	1.32	0.2879	
BC	5.62	1	5.62	0.036	0.8548	
A²	90.65	1	90.65	0.58	0.4705	
B²	97.52	1	97.52	0.63	0.4549	
C²	2245.32	1	2245.32	14.41	0.0068	
Residual	1090.77	7	155.82			
Lack of Fit	725.40	3	241.80	2.65	0.1852	not significant
Pure Error	365.37	4	91.34			
Cor Total	37469.02	16				

Std. Dev.	12.48	R-Squared	0.9709
Mean	297.56	Adj R-Squared	0.9335
C.V. %	4.20	Pred R-Squared	0.6750
PRESS	12177.25	Adeq Precision	17.052

Figure 3. Generated and printed a report regarding the ANOVA for total phenolic content. The report has been generated in Design-Expert v.7 Trial.

Factor	Coefficient Estimate	df	Standard Error	95% CI Low	95% CI High	VIF
Intercept	282.24	1	5.58	269.04	295.44	
A-Temperature	64.64	1	4.41	54.21	75.08	1.00
B-Time	-0.50	1	4.41	-10.94	9.93	1.00
C-Ultrasonic power	-2.89	1	4.41	-13.32	7.55	1.00
AB	5.12	1	6.24	-9.64	19.88	1.00
AC	-7.18	1	6.24	-21.94	7.58	1.00
BC	1.19	1	6.24	-13.57	15.94	1.00
A^2	4.64	1	6.08	-9.75	19.03	1.01
B^2	4.81	1	6.08	-9.57	19.20	1.01
C^2	23.09	1	6.08	8.71	37.48	1.01

Final Equation in Terms of Coded Factors:

TP =
+282.24
+64.64 * A
-0.50 * B
-2.89 * C
+5.12 * A * B
-7.18 * A * C
+1.19 * B * C
+4.64 * A^2
+4.81 * B^2
+23.09 * C^2

Final Equation in Terms of Actual Factors:

TP =
+282.24000
+64.64500 * Temperature
-0.50125 * Time
-2.88875 * Ultrasonic power
+5.12250 * Temperature * Time
-7.17750 * Temperature * Ultrasonic power
+1.18500 * Time * Ultrasonic power
+4.64000 * $Temperature^2$
+4.81250 * $Time^2$
+23.09250 * $Ultrasonic\ power^2$

Figure 4. Continuation of ANOVA report for total phenolic content generated in Design-Expert v.7 Trial.

In Figure 4 are shown estimated coefficients for TPC. As mentioned above, the only the linear term of temperature and quadratic term of ultrasonic power showed significant influence on TPC. Temperature also exhibits strong and positive influence, which is rather expected. Temperature increases diffusion and thus affects the mass transfer process. Increase in temperature causes degradation of plant matrix, which further improves solvent properties (Ramić et al., 2015).

Results presented in Figure 4 revealed the negative influence of the linear term of ultrasonic power and positive influence of its quadratic term. This indicated that TPC will decrease with increasing in ultrasonic power upto certain value after what TPC will slightly increase. In the end, the predicted model may be expressed with the following equation:

$$TPC = 282.24 + 64.64X_1 - 0.50X_2 - 2.89X_3 + 5.12X_1X_2 - 7.18X_1X_3 + 1.19X_2X_3 + 4.64X_1^2 + 4.81X_2^2 + 23.09X_3^2 \qquad (2)$$

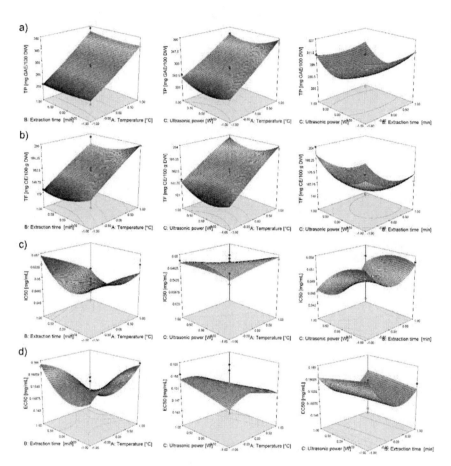

Figure 5. RSM plots showing the combined influence of UAE parameters on TPC (a), TFC (b), IC$_{50}$ (c) and EC$_{50}$ (d) (Zekovic et al., 2016).

Generally, all quadratic terms exhibit positive influence, while linear terms of extraction time and ultrasonic power showed a negative influence. Interactive coefficients are mostly positive with the exception of the interaction of temperature and ultrasonic power. Generated 3D plots (Figure

5) showed the influence of independent variables on all investigated parameters given in Table 3.

RSM approach has been also used for the optimization of SWE and SFE (Zeković, Pavlić, et al., 2016; Zeković et al., 2014). Zekovic et al., (Zeković et al., 2014) have optimized SWE using a similar approach, i.e., BBD only with 15 randomized experiments and 3 replicates in central point. Investigated responses are TPC, TFC and IC_{50}, like in the case of UAE and MAE, with the exception of the reduction power test. In this case, independent variables are temperature, pressure and extraction time. In this case, temperature and pressure have been the logical choice for optimization since these parameters exhibit the strongest influence on SWE process, which has been explained in the previous section.

Table 5. Experimental results for optimization of SFE using BBD design

Independent variable			Measured response*
X_1	X_2	X_3	Y (g/100 g)
0	0	0	3.77
0	-1	1	5.64
1	-1	0	5.95
0	0	0	4.00
0	1	1	3.50
-1	1	0	0.59
1	0	1	7.00
-1	-1	0	2.69
-1	0	1	1.20
0	0	0	4.02
0	-1	-1	4.31
1	0	-1	4.90
0	1	-1	2.05
-1	0	-1	0.95
1	1	0	5.36

*Results from Zeković et al., (2016).

Zekovic et al., (Zeković, Pavlić, et al., 2016) have also optimized SFE extraction. BBD model with 15 randomized experiments has been used with

3 replicates in central point. In this case, pressure and temperature have been selected as the main independent variables since they have the most significant influence on the process. The third variable is CO_2 flow rate, while the measured response is total extraction yield (expressed as grams per 100 g of coriander seed, g/100 g). Pressure ranges from 100 to 200 bar, temperature from 40 to 70°C and gas flow between 0.2 and 0.4 kg/h).

Table 5 shows that Y ranges from 0.59 to 7.00 g/100 g. The highest Y has been achieved at 200 bar, 55°C and 0.4 kg/h CO_2, while the lowest Y is noticed at 100 bar, 70°C and 0.3 kg/h. Experimental results are processed in the Design-Expert v.7 Trial software and reports of ANOVA are generated as in the case of UAE and MAE. Reports are presented in Figures 6 and 7.

Response 1 Yield
ANOVA for Response Surface Quadratic Model
Analysis of variance table [Partial sum of squares - Type III]

Source	Sum of Squares	df	Mean Square	F Value	p-value Prob > F	
Model	51.00	9	5.67	35.70	0.0005	significant
A-Pressure	39.56	1	39.56	249.25	< 0.0001	
B-Temperature	6.27	1	6.27	39.53	0.0015	
C-CO2 flow	3.31	1	3.31	20.83	0.0060	
AB	0.57	1	0.57	3.61	0.1159	
AC	0.86	1	0.86	5.41	0.0675	
BC	3.982E-003	1	3.982E-003	0.025	0.8804	
A^2	0.38	1	0.38	2.42	0.1803	
B^2	5.326E-003	1	5.326E-003	0.034	0.8619	
C^2	0.034	1	0.034	0.22	0.6607	
Residual	0.79	5	0.16			
Lack of Fit	0.76	3	0.25	13.38	0.0703	not significant
Pure Error	0.038	2	0.019			
Cor Total	51.80	14				

Std. Dev.	0.40	R-Squared	0.9847	
Mean	3.73	Adj R-Squared	0.9571	
C.V. %	10.68	Pred R-Squared	0.7648	
PRESS	12.18	Adeq Precision	20.545	

Figure 6. Generated report regarding the ANOVA for total extraction yield. The report has been generated in Design-Expert v.7 Trial.

Factor	Coefficient Estimate	df	Standard Error	95% CI Low	95% CI High	VIF
Intercept	3.93	1	0.23	3.34	4.52	
A-Pressure	2.22	1	0.14	1.86	2.59	1.00
B-Temperature	-0.89	1	0.14	-1.25	-0.52	1.00
C-CO2 flow	0.64	1	0.14	0.28	1.00	1.00
AB	0.38	1	0.20	-0.13	0.89	1.00
AC	0.46	1	0.20	-0.049	0.98	1.00
BC	0.032	1	0.20	-0.48	0.54	1.00
A²	-0.32	1	0.21	-0.86	0.21	1.01
B²	0.038	1	0.21	-0.50	0.57	1.01
C²	-0.097	1	0.21	-0.63	0.44	1.01

Final Equation in Terms of Coded Factors:

Yield =
+3.93
+2.22 * A
-0.89 * B
+0.64 * C
+0.38 * A*B
+0.46 * A*C
+0.032 * B*C
-0.32 * A²
+0.038 * B²
-0.097 * C²

Final Equation in Terms of Actual Factors:

Yield =
+3.93279
+2.22387 * Pressure
-0.88564 * Temperature
+0.64289 * CO2 flow
+0.37841 * Pressure * Temperature
+0.46337 * Pressure * CO2 flow
+0.031550 * Temperature * CO2 flow
-0.32276 * Pressure²
+0.037981 * Temperature²
-0.096644 * CO2 flow²

Figure 7. Continuation of ANOVA report for total extraction yield generated in Design-Expert v.7 Trial.

Report presented in Figure 6 showed that model is significant ($p < 0.05$), while lack of fit is insignificant ($p > 0.05$). Coefficient of multiple regression is in this case particularly high ($R^2 = 0.9847$), while the coefficient of variance is slightly higher than those in the case of MAE and UAE ($CV = 10.68\%$).

The report also showed that all three linear terms exhibit significant influence on the model. This is rather expected especially in the case of pressure and temperature due to their strong influence on SFE. Estimated coefficients (Figure 7) revealed a strong and positive effect of pressure on SFE, while the quadratic term is low and negative. On the other hand, the linear term of temperature showed negative influence, while the quadratic term is lower and positive. This indicates that Y will increase with increasing pressure and decreasing temperature. Such influence of pressure is expected since pressure influences positively on the density of carbon dioxide, which results in elevated solubility of compounds in supercritical fluid (Paixao Coelho & Figueiredo Palavra, 2015; Pourmortazavi & Hajimirsadeghi, 2007). The temperature in this case shows more complex than the influence of pressure. Temperature influences on the density of the supercritical fluid and on vapour pressure. With the increase in temperature, the density of carbon-dioxide decreases, while vapour pressure increases. The decrease in density causes lower solubility, but the increase in vapour pressure results in increased solubility (Jesus & Meireles, 2014; Paixao Coelho & Figueiredo Palavra, 2015). In this particular case, the influence of temperature on density is the dominant phenomenon, thus Y decreases.

Estimated coefficients showed a positive influence of the linear term of carbon-dioxide flow. This indicates that yield will increase with the flow. A possible explanation for such results is the reduced thickness of the film around the solid particles. This results in a decrease of resistance and better mass transfer (Döker, Salgın, Şanal, Mehmetoğlu, & Çalımlı, 2004). Using equation (1) polynomial model for Y is generated:

$$Y = 3.93 + 2.22X_1 - 0.89X_2 + 0.64X_3 + 0.38X_1X_2 + 0.46X_1X_3 + 0.032X_2X_3 - 0.32X_1^2 + 0.038X_2^2 - 0.097X_3^2 \qquad (3)$$

3D models of the combined influence of described parameters on total extraction yield are given in Figure 8. Presented plots clearly showed a positive influence of pressure and carbon-dioxide flow on Y and negative influence of temperature on the same parameter.

Another approach which has been also widely applied for optimization of the process is an artificial neural network (ANN). This method acts like a natural neural system with the help of a computer. Advantages of this approach are non-linearity, adaptively, generalization, model independence, easy to use and high accuracy. Its connection weights are used for determination of relative importance of various inputs (Gevrey, Dimopoulos, & Lek, 2003; Olden & Jackson, 2002; Tchaban, Taylor, & Griffin, 1998; Yoon, Swales, & Margavio, 1993). ANN has been successfully applied for prediction of SFE extraction kinetics and for optimization of the process (Azmir et al., 2014; Khajeh, Moghaddam, & Shakeri, 2012; Kuvendziev, Lisichkov, Zeković, & Marinkovski, 2014; Shokri, Hatami, & Khamforoush, 2011; Sodeifian, Sajadian, & Saadati Ardestani, 2016; Zahedi & Azarpour, 2011).

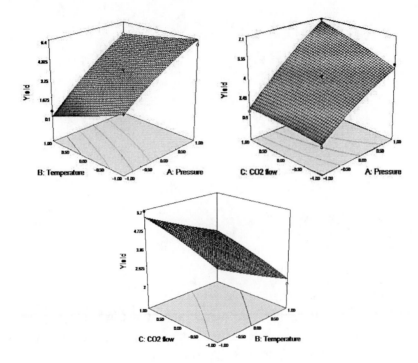

Figure 8. RSM plots showing the combined influence of SFE parameters on Y (Zeković, Pavlić, et al., 2016).

Model for optimization of the SFE process for coriander seeds extraction has been developed in MATLAB software using MATLAB Neural Network Toolbox. The model has three inputs (pressure, temperature and carbon-dioxide flow rate) and one output (initial slope obtained from kinetics model). Schematic model is given in Figure 9. In the end, the model has been used for the optimization of the SFE process like in the case of RSM. In this case, calculation of extraction kinetics parameters is combined with the optimization process in order to achieve maximal Y for shorter possible time. The calculated initial slope has been used as a parameter which represents the initial phase of the extraction process, i.e., solubility-controlled phase. Since extraction time is 4 h initial slope has been selected as the response variable in this case (Zeković, Bera, et al., 2017).

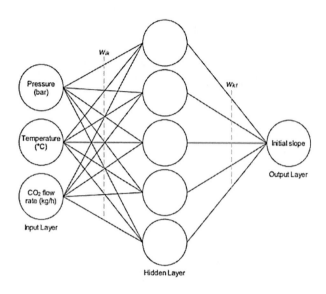

Figure 9. Schematic representation of ANN architecture.

ANN results, which includes weights values, varies with changes in initial values used for ANN construction and fitting. Different numbers of hidden neurons may also give different results of optimization. To avoid these impacts on final results, a number of neurons in the hidden layer have been varied from 1 to 20, while training process of each network is repeated 10 times with random initial values of weights and biases. As the final results, 200 ANNs have been created of which 194 are used for further analysis ($R^2 > 0.8$). Mean value for R^2 is 0.932, while best fitting has been achieved with 5 hidden neurons ($R^2 = 0.979$) (Zeković, Bera, et al., 2017). Influence of hidden neuron on fitting value after 10 repeated trainings is shown in Figure 10.

Figure 10 shows that neurons positively influence to R^2 mean value, i.e., increasing in neurons in hidden layer causes an increase into R^2 mean value, which is always higher than 0.9. The best of 200 ANNs with 5 hidden neurons are used for further optimization calculations in order to avoid "overfitting" with a high number of hidden neurons. Investigation of the influence of hidden neurons on relative importance (RI) mean values after 10 pieces of training is presented in Figure 11a, while mean values of all calculated RI with standard deviation is given in Figure 11b.

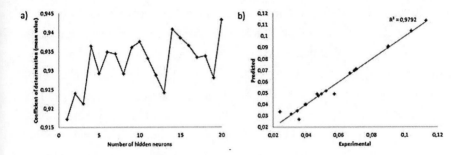

Figure 10. Influence of hidden neurons on the coefficient of determination (a) and regression plot for ANN with 5 hidden neurons (Zeković, Bera, et al., 2017).

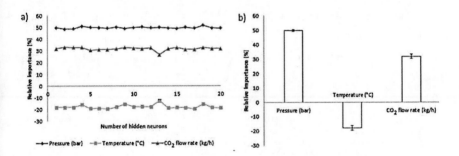

Figure 11. Influence of the number of hidden neurons on RI mean values (a) and mean values of all calculated RIs with standard deviation (b) (Zeković, Bera, et al., 2017).

Table 6. Optimized SFE parameters using different optimization approaches

Optimization process	Pressure (bar)	Temperature (°C)	Flow rate (kg/h)
RSM*	199.50	40.15	0.396
ANN**	200	40	0.4

*Results from Zeković et al., 2016 (Zeković, Pavlić, et al., 2016), **Results from Zeković et al., 2017 (Zeković, Bera, et al., 2017).

Figure 11 showed that pressure is the most influential parameter with about 50% of relative importance. On the other hand, temperature and carbon-dioxide flow rate showed about 18% and 32% of relative importance, respectively. it might be also noticed that pressure and flow rate exhibit positive influence, while temperature showed a negative one. This is in agreement with results obtained by RSM. The positive influence of pressure in initial slope which may be explained with increasing density of carbon-dioxide and thus increasing in extraction rate. The positive influence of flow rate may be explained by the fact that initial phase of extraction depends on on solubility of desired compounds and continual import of fresh solvent, this provides faster dissolution of compounds due to high concentration gradient from solid to carbon-dioxide. The negative influence of temperature is caused by a decrease of density of carbon dioxide with increase temperature. Optimized values of parameters obtained by RSM and ANN are shown in Table 6.

From the presented results in Table 6, it may be concluded that both optimization approaches give similar results. This indicates that both tools are suitable for optimization of supercritical fluid extraction of coriander seeds.

CHEMICAL COMPOSITION OF CORIANDER EXTRACTS

All available studies showed that coriander is of interest for the scientific community due to its essential oils. Thus, most studies have dealt with isolation and characterization of essential oils of coriander. Supercritical fluid extraction has been the most applied technique in this case besides hydrodistillation. All conducted studies are agreed that linalool is main constituent of essential oil (>50%) (Alves-Silva et al., 2013; Baratta, Dorman, Deans, Biondi, & Ruberto, 1998; Gil et al., 2002; Mandal & Mandal, 2015; Mhemdi et al., 2011; Msaada et al., 2007; Msaada, Taarit, Hosni, Hammami, & Marzouk, 2009; Ravi, Prakash, & Bhat, 2007; Sahib et al., 2012; Shahwar et al., 2012; Singh, Maurya, de Lampasona, & Catalan, 2006; Sourmaghi, Kiaee, Golfakhrabadi, Jamalifar, & Khanavi, 2015; Sriti,

Talou, Wannes, Cerny, & Marzouk, 2009; Sriti, Wannes, Talou, Vilarem, & Marzouk, 2011; Teixeira et al., 2013; Zoubiri & Baaliouamer, 2010). Beside linalool, geraniol, camphor and limonene are also presented in significant amount (Pavlić et al., 2015; Zeković et al., 2015).

Generally, presented terpenes may be divided into several groups: hemiterpenes, monoterpenes, sesquiterpenes, diterpenes, sesterterpenes, triterpenes, tetraterpenes and polyterpenes. Monoterpenes (two isoprene units, 10 carbon atoms) may be further divided into acyclic monoterpenes, cyclic monoterpenes, acyclic oxygenated monoterpenes, cyclic oxygenated monoterpenes and aromatic oxygenated monoterpenes. Presence of several subgroups of monoterpenes have been reported (Table 7) (Zeković, Pavlić, et al., 2016) It might be noticed that the composition of extracts is quite different. Parameters of SFE extraction obviously influence on the solubility of compounds in supercritical carbon dioxide. The difference among these three samples is clearly noticeable when comparing their chromatograms.

Structures of most common compounds in coriander seeds extracts are presented in Figure 13. Diversity in their structure is the main reason for different solubility in the supercritical fluid.

Limonene and γ-terpinene have same molecular formula ($C_{10}H_{16}$). On the other hand, geraniol, α-terpineol, eucalyptol and linalool also have the same formula ($C_{10}H_{18}O$). A closer look at their structures shows significant differences. Limonene and γ-terpinene differ in the position of double bonds. Geraniol and linalool possess hydroxyl group in a different position. Geraniol is primary alcohol, while linalool is tertiary alcohol. Results presented by Zeković et al., (Zeković, Pavlić, et al., 2016) showed that geraniol is better soluble at a higher temperature, while linalool is better soluble at a lower temperature. Such behaviour of those two alcohols may be explained by the occurrence of self-association among molecules (Tufeu, Subra, & Plateaux, 1993). Alcohols with increased branching possess lower self-associative affinity thus solubility increases (Friedrich & Schneider, 1989). Consequently, linalool is better soluble at a lower temperature, while the higher temperature is needed to extract geraniol. Same explanation goes for α-terpineol which is also tertiary alcohol with cyclic system and a double bond.

Table 7. Chemical composition of SFE extracts of coriander seeds

Compound	Sample*		
	3	11	15
Cyclic monoterpenes			
(+)-Limonene	2.0	1.0	< 0.1
α-Pinene	< 0.1	< 0.1	< 0.1
β-Pinene	< 0.1	< 0.1	< 0.1
γ-Terpinene	3.0	< 0.1	2.1
p-Cymene	D**	D	D
Acyclic oxygenated monoterpenes			
Geraniol	3.0	5.2	10.7
Geranyl acetate	D	ND***	ND
Linalool	717.0	642.0	608.2
Nerol	D	ND	ND
Cyclic oxygenated monoterpenes			
Camphor	21.0	10.9	9.6
Eucalyptol	< 0.1	< 0.1	1.2
Terpinen-4-ol	D**	ND***	ND
α-Terpineol	2.0	6.1	4.2
Aromatic oxygenated monoterpenes			
Carvacrol	< 0.1	< 0.1	< 0.1
Eugenol	< 0.1	2.0	20.7
Methyl cavicol	1.0	18.2	13.2
Furanoids			
cis-Linalool oxide	D	D	D
trans-Linalool oxide	D	D	D

*Results from Zeković et al., (2016), **D-detected, ***ND-not detected.

Camphor is ketone, while eucalyptol possesses an etheric functional group. Both molecules are cyclic systems. Aldehyde and ketones are generally soluble in the supercritical fluid. On the other hand, eucalyptol is a saturated system which decreases its solubility in carbon-dioxide (Dandge, Heller, & Wilson, 1985). Eugenol and methyl chavicol are aromatic systems and both possess propenyl and methoxy groups, while eugenol has an additional hydroxy functional group. Presence of methoxy and hydrocarbon groups increase solubility in carbon-dioxide (Dandge et al., 1985). Additional hydroxy group modifies the solubility of eugenol toward the higher temperatures, which may be explained with mentioned self-association of molecules (Friedrich & Schneider, 1989; Tufeu et al., 1993).

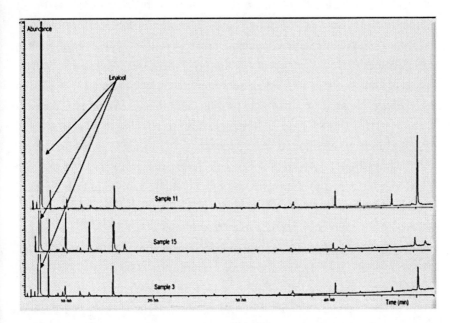

Figure 12. Chromatograms of samples no. 3, 11 and 15 (Zeković, Pavlić, et al., 2016).

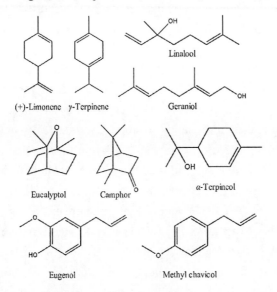

Figure 13. Structures of most common monoterpenes in coriander seeds extracts.

The leaf has been also used for isolation of essential oil. Presence of different aldehydes and alcohols. 2E-decenal (15.9%), decanal (14.3%), 2E-decen-1-ol (14.2%) and n-decanol (13.6%) have been reported as major compounds in leaves essential oil. 2E-tridecen-1-al, 2E-dodecenal, dodecanal, undecanol and undecanal are reported as minor compounds, while alkanes (1.46%) represent remaining compounds (Matasyoh, Maiyo, Ngure, & Chepkorir, 2009). The essential oil from the coriander leaves from Bangladesh showed a different profile. Dominant compounds are 2-decenoic acid (30.8%), E-11-tetradecenoic acid (13.4%), capric acid (12.7%), undecyl alcohol (6.4%), tridecanoic acid (5.5%) and undecanoic acid (7.1%) (Bhuiyan, Begum, & Sultana, 2009). Generally, the composition of oil varied from the region to region, but main compounds in leaves' essential oil are aldehydes and alcohols, while linalool dominates in seeds' essential oil (Sahib et al., 2012). Beside regional differences, the chemical composition of plants extracts are dependent on the maturity of the plant (Msaada et al., 2007), environmental conditions (Gil et al., 2002) and applied extraction technique (Pavlić et al., 2015; Sourmaghi et al., 2015).

Besides terpenes, aldehyde and alcohols, fatty acids and sterols have also been found in plant material (Mhemdi et al., 2011; Singh et al., 2006; Sriti et al., 2009). Singh et al., (Singh et al., 2006) reported oleic acid as the main fatty acid (36.52%) followed by linoleic acid (33.21%). Sriti et al., (2009) have analyzed fatty acids and sterols in seed, pericarp and whole fruit of coriander. Authors have reported petroselinic acid as the main compound with contents of 76.37%, 42.20% and 75.07% in seed, pericarp and whole fruit of coriander, respectively. Higher content has been achieved by linoleic, oleic and palmitic acids. In total, saturated fatty acids have been presented in 4.44%, 27.97% and 5.17%, while unsaturated acid achieved 95.20%, 72.03% and 94.82% in seed, pericarp and whole fruit of coriander, respectively. it might notice that seed and fruits are richer in unsaturated acids than pericarp. On the other side, pericarp has a higher content of saturated acids. (Mhemdi et al., (2011) investigated monounsaturated and polyunsaturated fatty acids separately. Authors have confirmed that unsaturated acids are more prevalent than saturated, but monounsaturated showed higher content than polyunsaturated acids. Content of saturated acids varied from 4.59% to 5.64%, monounsaturated in the range of 78.8-81.4% and polyunsaturated varied from 14.0% to 15.5%. The most abundant saturated acid is palmitic acid (3.45-4.24%), monounsaturated acid is petroselinic acid (72.3-75.3%) and polyunsaturated acid is linoleic acid (13.7-15.3%) (Mhemdi et al., 2011).

As far as sterols are concerned, Sriti et al., (Sriti et al., 2009) have reported the presence of cholesterol, campesterol, stigmasterol, β-sitosterol, Δ^5-avenasterol, Δ^5, 24-stigmastadienol, Δ^7-stigmasterol and Δ^7-avenasterol. These compounds have been detected in all analyzed parts of the plant, but stigmasterol is the most abundant in seed, β-sitosterol in pericarp and whole fruit. It should be mentioned that Δ^5, 24-stigmastadienol is detected in pericarp in a higher amount, while both stigmasterol and β-sitosterol dominate in seed and fruit followed by Δ^7-stigmasterol (Sriti et al., 2009).

BIOLOGICAL ACTIVITY OF CORIANDER

The plant is usually used as food and in pharmaceutical and cosmetic industries (Aluko, McIntosh, & Reaney, 2001; Eyres, Dufour, Hallifax, Sotheeswaran, & Marriott, 2005; Jabeen, Bashir, Lyoussi, & Gilani, 2009; Mhemdi, Rodier, Kechaou, & Fages, 2011). Many different studies showed that coriander and its extracts express various activities such as antioxidant, cytotoxic, antimicrobial, anticarcinogenic, antimutagenic and and antidiabetic activity (Chen et al., 2009; Chithra & Leelamma, 2000; Delaquis, 2002; Gallagher, Flatt, Duffy, & Abdel-Wahab, 2003; Grosso et al., 2008; Lindberg Madsen, 1995; Lo Cantore, Iacobellis, De Marco, Capasso, & Senatore, 2004; Ramadan, Kroh, & Mörsel, 2003; Wangensteen et al., 2004). Such a wide range of activity of seeds possesses by virtue of its chemical composition.

Coriander and its parts and products have been used across the world in traditional medicine. The most commonly applied parts and products are leaves, seed and essential oils. Leaves are known by its antispasmodic, dyspeptic and appetizer effects. It is also used to treat abdominal discomfort. Preparations made from leaves have been used for the treatment of coughs, chest pain, bladder complaints and also as an aphrodisiac. Coriander fruit has been used for the treatment of inflammation, indigestion, cough, bronchitis, vomiting, dysentery, diarrhoea, gout, rheumatism, etc (Sahib et al., 2012).

Coriander, its extracts and essential oil have shown a wide range of biological activity. One of the most investigated one is antimicrobial activity. Investigations have shown that essential oil and different extracts have shown activity against many bacteria and yeast species (Begnami, Duarte, Furletti, & Rehder, 2010; Kubo, Fujita, Kubo, Nihei, & Ogura, 2004; Matasyoh et al., 2009). The essential oil has shown activity against both Gram-positive and Gram-negative bacteria, as well as antifungal activity against *Candida albicans*. Leaf essential oils also inhibit a number of Candida species in the concentration range of 125-500 μg/mL (Sahib et al., 2012). Antifungal and sprout suppressant activities have been also studied and proved. In this case, essential oils showed higher activity than oleoresin (Singh et al., 2006). Investigations have shown that aqueous and methanolic extracts from leaves have induced cell damage and thus inhibit the growth of microbes. Methanolic extracts show higher activity, while stem extracts possess higher TPC and higher activity than leaves extracts indicating a correlation between TPC and antimicrobial activity (Wong & Kitts, 2006). Antibacterial activity of essential oils has been attributed to the presence of alcohols and aldehydes and other compounds such as α-pinene, camphene and linalool. These compounds have proved to be active against a wide range of both Gram-positive and Gram-negative bacteria (Delaquis, 2002).

Antioxidant activity of coriander essential oils and other extracts is widely investigated and well known (Baratta et al., 1998; Gallo et al., 2010; Wangensteen et al., 2004; Zeković, Bera, et al., 2017; Zekovic et al., 2016; Zeković, Pavlić, et al., 2016; Zeković et al., 2014; Zeković, Vladić, et al., 2016). Wangensteen et al., (Wangensteen et al., 2004) have investigated the antioxidant activity of extracts with different polarity and of oil itself. Extracts obtained with ethyl acetate (medium polarity) has shown strong activity in the case of the DPPH test. They have concluded that leaf extracts have higher activity than seed extract. Another study has shown that leaf methanolic and aqueous extract possesses higher activity against free radical than stem extract obtained with the same solvents (Wong & Kitts, 2006). Hashim et al., (Hashim, Lincy, Remya, Teena, & Anila, 2005) have investigated the activity of coriander extract on H_2O_2-induced stress in human lymphocytes. Treatment with phenolic fraction in the concentration of 50 μg/mL efficiently protected cells from oxidative stress by restoring it to a normal level. There is also evidence that coriander has a hepatoprotective effect against CCl_4-induced toxicity (Pandey, Bigoniya, Raj, & Patel, 2011). These effects of the extracts have been attributed to the high content of iso-quercetin and quercetin. Antioxidant activity has been also a subject of optimization process (Zekovic et al., 2016; Zeković, Pavlić, et al., 2016; Zeković et al., 2014; Zeković, Vladić, et al., 2016). RSM has been employed to obtain extract with the highest antioxidant activity using different extraction techniques such as UAE (Zekovic et al., 2016), MAE (Zeković, Vladić, et al., 2016), SWE (Zeković et al., 2014) and SFE (Zeković, Pavlić, et al., 2016). Although all these techniques imply the application of different solvent with different polarity, all of them showed significant antioxidant activity.

As previously mentioned, coriander has been used as an antidiabetic agent. Application of coriander has shown to increase glucose uptake, glucose oxidation, glycogenesis and dose-dependent effect on insulin secretion (Gray & Flatt, 1999). Decreasing in serum glucose concentration and increase in activity of beta cells have been observed after addition of 200-250 mg/kg so ethanolic extract in comparison with diabetic control (Eidi et al., 2009). Aqueous extract of coriander seeds has suppressed hyperglycemia, while normal glucose level has been reached after 4 h of dosing (Sahib et al., 2012). Chithra and Leelamma (Chithra & Leelamma, 1999) have shown that pretreatment with coriander seed powder causes certain changes in the metabolism of carbohydrates. Increasing in the concentration of hepatic glycogen and glycogen synthase has been noticed. Beside hyperglycemic effect, coriander shows antidyslipidemic activity. It is showing lipid-lowering effect. The decrease in triglyceride level, LDL and VLDL, as well as an increase in HDL have been observed (Chithra & Leelamma, 1997).

Beside mentioned activities, coriander and its extracts have been tested for other activities such as anticonvulsant, anxiolytic, sedative, antidepressant, cognitive, antimutagenic, diuretic, antihypertensive, anti-inflammatory and miscellaneous effects (Sahib et al., 2012) and studies regarding this plant is still going on.

CONCLUSION

Coriander has been used in various civilizations and nations during a long period of time as food and medicine. This plant is still significant merchandise and is in focus of the scientific community. Various parts of the plant have been investigated in order to determine chemical composition. In order to do so, different extraction approaches have been applied. The most investigated part of the plant is its seed, mainly for essential oil, which is widely used in the world. The investigation has shown that monoterpene linalool is the main compound, followed by geraniol, camphor and limonene. Different solvents have been used for extraction of compounds from coriander ranging from polar to nonpolar. Beside chemical composition, the biological activity of extracts has been assessed. Obtained results showed that plant and its extracts express a wide range of activity, which justifies its application in folk medicine in many different countries. Studies have also shown that chemical composition and activity is in strong correlation. Aiming to obtain extracts with a maximal concentration of biologically active compounds and maximal activity, response surface methodology and artificial neural network have been applied to optimize extraction processes. All in all, results have shown that this plant justifiably possesses one of the most significant places in pharmaceutical and food industries in these days.

REFERENCES

Adamczyk J, Horny N, Tricoteaux A, Jouan, PY, Zadam M (2008). On the use of response surface methodology to predict and interpret the preferred c-axis orientation of sputtered AlN thin films. *Applied Surface Science*, 254, 1744–1750.

Aluko RE, McIntosh T, Reaney M (2001). Comparative study of the emulsifying and foaming properties of defatted coriander (*Coriandrum sativum*) seed flour and protein concentrate. *Food Research International*, 34, 733–738.

Alves-Silva JM, Dias dos Santos SM, Pintado ME, Pérez-Álvarez JA, Fernández-López J, Viuda-Martos M (2013). Chemical composition and in vitro antimicrobial, antifungal and antioxidant properties of essential oils obtained from some herbs widely used in Portugal. *Food Control*, 32, 371–378.

Azmir J, Zaidul ISM, Rahman MM, Sharif KM, Mohamed A, Sahena F, Jahurul JHA, Ghafoor K, Norulaini NAN, Omar AKM (2013). Techniques for extraction of bioactive compounds from plant materials: A review. *Journal of Food Engineering*, 117, 426–436.

Azmir J, Zaidul ISM, Sharif KM, Uddin MS, Jahurul MHA, Jinap S, Hajeb P, Mohamed A (2014). Supercritical carbon dioxide extraction of highly unsaturated oil from Phaleria macrocarpa seed. *Food Research International*, 65, 394–400.

Baratta MT, Dorman HJD, Deans SG, Biondi DM, Ruberto G (1998). Chemical Composition, Antimicrobial and Antioxidative Activity of Laurel, Sage, Rosemary, Oregano and Coriander Essential Oils. *Journal of Essential Oil Research*, 10, 618–627.

Baş D, Boyacı İH (2007). Modeling and optimization I: Usability of response surface methodology. *Journal of Food Engineering*, 78, 836–845.

Begnami AF, Duarte MCT, Furletti V, Rehder VLG (2010). Antimicrobial potential of *Coriandrum sativum* L. against different *Candida* species in vitro. *Food Chemistry*, 118, 74–77.

Bhuiyan MNI, Begum J, Sultana M (2009). Chemical composition of leaf and seed essential oil of *Coriandrum sativum* L. from Bangladesh. *Bangladesh Journal of Pharmacology*, 4, 150-153.

Brunner G (1994) *Gas Extraction: An Introduction to Fundamentals of Supercritical Fluids and the Application to Separation Processes* (Vol. 4). Heidelberg: Steinkopff.

Brunner G (2005). Supercritical fluids: technology and application to food processing. *Journal of Food Engineering*, 67, 21–33.

Chemat F, Tomao V, Virot M (2008). Ultrasound-Assisted Extraction in Food Analysis. In Semih Ötles (Ed.), *Handbook of Food Analysis Instruments* (pp. 85–94). Boca Raton, Florida, USA: CRC press.

Chen Q, Yao S, Huang X, Luo J, Wang J, Kong L (2009). Supercritical fluid extraction of *Coriandrum sativum* and subsequent separation of isocoumarins by high-speed counter-current chromatography. *Food Chemistry*, 117, 504–508.

Chithra V, Leelamma S (1997). Hypolipidemic effect of coriander seeds (*Coriandrum sativum*): mechanism of action. *Plant Foods for Human Nutrition* (Dordrecht, Netherlands), 51, 167–172.

Chithra V, Leelamma S (1999). *Coriandrum sativum* — mechanism of hypoglycemic action. *Food Chemistry*, 67, 229–231.

Chithra V, Leelamma S (2000). *Coriandrum sativum* — effect on lipid metabolism in 1,2-dimethyl hydrazine induced colon cancer. *Journal of Ethnopharmacology*, 71, 457–463.

Coşkuner Y, Karababa E (2007). Physical properties of coriander seeds (*Coriandrum sativum* L.). *Journal of Food Engineering*, 80, 408–416.

Cravotto G, Boffa L, Mantegna S, Perego P, Avogadro M, Cintas P (2008). Improved extraction of vegetable oils under high-intensity ultrasound and/or microwaves. *Ultrasonics Sonochemistry*, 15, 898–902.

Cvetanović A, Švarc-Gajić J, Gašić U, Tešić Ž, Zengin G, Zeković Z, Đurović S (2016). Isolation of apigenin from subcritical water extracts: Optimization of the process. *The Journal of Supercritical Fluids*, 120, 32–42.

Cvetanović A, Švarc-Gajić J, Zeković Z, Gašić U, Tešić Ž, Zengin G, Mašković P, Mahomoodally, MF Đurović, S. (2018). Subcritical water extraction as a cutting edge technology for the extraction of bioactive compounds from chamomile: Influence of pressure on chemical composition and bioactivity of extracts. *Food Chemistry*, 266, 389–396.

Cvetanović A, Švarc-Gajić J, Zeković Z, Jerković J, Zengin G, Gašić U, Tešić Ž, Mašković P, Đurović S (2019). The influence of the extraction temperature on polyphenolic profiles and bioactivity of chamomile (*Matricaria chamomilla* L.) subcritical water extracts. *Food Chemistry*, 271, 328–337.

Cvetanović A, Švarc-Gajić J, Zeković Z, Mašković P, Đurović S, Zengin G, Delerue-Matos C, Lozano-Sánchez J, Jakšić A (2017). Chemical and biological insights on aronia stems extracts obtained by different extraction techniques: From wastes to functional products. *The Journal of Supercritical Fluids*, 128, 173–181.

da Silva RPFF, Rocha-Santos TAP, Duarte AC (2016). Supercritical fluid extraction of bioactive compounds. *TrAC Trends in Analytical Chemistry*, 76, 40–51.

Dandge DK, Heller JP, Wilson KV (1985). Structure solubility correlations: organic compounds and dense carbon dioxide binary systems. *Industrial & Engineering Chemistry Product Research and Development*, 24, 162–166.

de Melo MMR, Silvestre AJD, Silva CM (2014). Supercritical fluid extraction of vegetable matrices: Applications, trends and future perspectives of a convincing green technology. *The Journal of Supercritical Fluids*, 92, 115–176.

Delaquis P (2002). Antimicrobial activity of individual and mixed fractions of dill, cilantro, coriander and eucalyptus essential oils. *International Journal of Food Microbiology*, 74, 101–109.

Dima C, Ifrim GA, Coman G, Alexe P, Dima Ş (2016). Supercritical CO_2 Extraction and Characterization of *Coriandrum sativum* L. Essential Oil. *Journal of Food Process Engineering*, 39, 204–211.

Döker O, Salgın U, Şanal İ, Mehmetoğlu Ü, Çalımlı A (2004). Modeling of extraction of β-carotene from apricot bagasse using supercritical CO_2 in packed bed extractor. *The Journal of Supercritical Fluids*, 28, 11–19.

Đurović S (2019). *Urtica dioica, contemporary extraction techniques, chemical profile, biological activity, formulation of food product*. PhD disertation, University of Novi Sad, Faculty of Technology.

Đurović S, Šorgić S, Popov S, Radojković M, Zeković Z (2018). Isolation and GC Analysis of Fatty Acids: Study Case of Stinging Nettle Leaves. In *Carboxylic Acid - Key Role in Life Sciences*. InTech.

Eidi M, Eidi A, Saeidi A, Molanaei S, Sadeghipour A, Bahar M, Bahar K (2009). Effect of coriander seed (*Coriandrum sativum* L.) ethanol extract on insulin release from pancreatic beta cells in streptozotocin-induced diabetic rats. *Phytotherapy Research*, 23, 404–406.

Eikani MH, Golmohammad F, Rowshanzamir S (2007). Subcritical water extraction of essential oils from coriander seeds (*Coriandrum sativum* L.). *Journal of Food Engineering*, 80, 735–740.

Eyres G, Dufour JP, Hallifax G, Sotheeswaran S, Marriott PJ (2005). Identification of character-impact odorants in coriander and wild coriander leaves using gas chromatography-olfactometry (GCO) and comprehensive two-dimensional gas chromatography-time-of-flight mass spectrometry (GC x GC-TOFMS). *Journal of Separation Science*, 28, 1061–1074.

Filip S, Vidović S, Vladić J, Pavlić B, Adamović D, Zeković Z (2016). Chemical composition and antioxidant properties of *Ocimum basilicum* L. extracts obtained by supercritical carbon dioxide extraction: Drug exhausting method. *The Journal of Supercritical Fluids*, 109, 20–25.

Friedrich J, Schneider GM (1989). Near-infrared spectroscopic investigations on phase behaviour and association of 1-octadecanol, 3-hexanol, and 3-methyl-3-pentanol in carbon dioxide and chlorotrifluoromethane. *The Journal of Chemical Thermodynamics*, 21, 307–319.

Gallagher A, Flatt P, Duffy G, Abdel-Wahab YH (2003). The effects of traditional antidiabetic plants on in vitro glucose diffusion. *Nutrition Research*, 23, 413–424.

Gallo M, Ferracane R, Graziani G, Ritieni A, Fogliano V (2010). Microwave Assisted Extraction of Phenolic Compounds from Four Different Spices. *Molecules*, 15, 6365–6374.

Gevrey M, Dimopoulos I, Lek S (2003). Review and comparison of methods to study the contribution of variables in artificial neural network models. *Ecological Modelling*, 160, 249–264.

Gil A, de la Fuente EB, Lenardis AE, López Pereira M, Suárez SA, Bandoni A, van Baron C, Di Leo Lira P, Ghersa CM (2002). Coriander Essential Oil Composition from Two Genotypes Grown in Different Environmental Conditions. *Journal of Agricultural and Food Chemistry*, 50, 2870–2877.

Gray AM, Flatt PR (1999). Insulin-releasing and insulin-like activity of the traditional anti-diabetic plant *Coriandrum sativum* (coriander). *The British Journal of Nutrition*, 81, 203–209.

Grosso C, Ferraro V, Figueiredo AC, Barroso JG, Coelho JA, Palavra AM (2008). Supercritical carbon dioxide extraction of volatile oil from Italian coriander seeds. *Food Chemistry*, 111, 197–203.

Hashim M, Lincy S, Remya V, Teena M, Anila L (2005). Effect of polyphenolic compounds from on HO-induced oxidative stress in human lymphocytes. *Food Chemistry*, 92, 653–660.

Herrero M, Cifuentes A, Ibanez E (2006). Sub- and supercritical fluid extraction of functional ingredients from different natural sources: Plants, food-by-products, algae and microalgae: A review. *Food Chemistry*, 98, 136–148.

Jabeen Q, Bashir S, Lyoussi B, Gilani AH (2009). Coriander fruit exhibits gut modulatory, blood pressure lowering and diuretic activities. *Journal of Ethnopharmacology*, 122, 123–130.

Jain T, Jain V, Pandey R, Vyas A, Shukla SS (2009). Microwave assisted extraction for phytoconstituents – An overview. *Asian Journal of Research in Chemistry*, 2, 19–25.

Jesus SP, Meireles MAM (2014). Supercritical fluid extraction: A global perspective of the fundamental concepts of this eco-friendly extraction technique. In F Chemat & MA Vian (Eds.), *Alternative solvents for natural products extraction* (pp. 39–72). Berlin: Springer-Verlag.

Khajeh M, Moghaddam MG, Shakeri M (2012). Application of artificial neural network in predicting the extraction yield of essential oils of

Diplotaenia cachrydifolia by supercritical fluid extraction. *The Journal of Supercritical Fluids*, 69, 91–96.

Ko MJ, Cheigh CI, Chung MS (2014). Relationship analysis between flavonoids structure and subcritical water extraction (SWE). *Food Chemistry*, 143, 147–155.

Kubo I, Fujita K, Kubo A, Nihei K, Ogura T (2004). Antibacterial Activity of Coriander Volatile Compounds against *Salmonella choleraesuis*. *Journal of Agricultural and Food Chemistry*, 52, 3329–3332.

Kuvendziev S, Lisichkov K, Zeković Z, Marinkovski M (2014). Artificial neural network modelling of supercritical fluid CO_2 extraction of polyunsaturated fatty acids from common carp (*Cyprinus carpio* L.) viscera. *The Journal of Supercritical Fluids*, 92, 242–248.

Lindberg Madsen H (1995). Spices as antioxidants. *Trends in Food Science & Technology*, 6, 271–277.

Lo Cantore P, Iacobellis NS, De Marco A, Capasso F, Senatore F (2004). Antibacterial Activity of *Coriandrum sativum* L. and *Foeniculum vulgare* Miller Var. *vulgare* (Miller) Essential Oils. *Journal of Agricultural and Food Chemistry*, 52, 7862–7866.

Mandal S, Mandal M (2015). Coriander (*Coriandrum sativum* L.) essential oil: Chemistry and biological activity. *Asian Pacific Journal of Tropical Biomedicine*, 5, 421–428.

Mašković P, Radojković M, Cvetanović A, Mitić M, Zeković Z, Đurović S (2018). Chemical profile and biological activity of tart cherry twigs : possibilities of plant waste utilization. *Journal of Food and Nutrition Research*, 57, 222–230.

Mašković P, Veličković V, Mitić M, Đurović S, Zeković Z, Radojković M, Cvetanović A, Švarc-Gajić J, Vujić J (2017). Summer savory extracts prepared by novel extraction methods resulted in enhanced biological activity. *Industrial Crops and Products*, 109, 875–881.

Mašković PZ, Veličković V, Đurović S, Zeković Z, Radojković M, Cvetanović, A, Mitić M, Zeković Z, Vujić J (2018). Biological activity and chemical profile of Lavatera thuringiaca L. extracts obtained by different extraction approaches. *Phytomedicine*, 38, 118–124.

Matasyoh JC, Maiyo ZC, Ngure RM, Chepkorir R (2009). Chemical composition and antimicrobial activity of the essential oil of *Coriandrum sativum*. *Food Chemistry*, 113, 526–529.

Meullemiestre A, Breil C, Abert-Vian M, Chemat F (2015). *Modern Techniques and Solvents for the Extraction of Microbial Oils*. Cham: Springer International Publishing.

Mhemdi H, Rodier E, Kechaou N, Fages J (2011). A supercritical tuneable process for the selective extraction of fats and essential oil from coriander seeds. *Journal of Food Engineering*, 105, 609–616.

Msaada K, Hosni K, Taarit M, Chahed T, Kchouk ME, Marzouk B (2007). Changes on essential oil composition of coriander (*Coriandrum sativum* L.) fruits during three stages of maturity. *Food Chemistry*, 102, 1131–1134.

Msaada K, Taarit M, Hosni K, Hammami M, Marzouk B (2009). Regional and maturational effects on essential oils yields and composition of coriander (*Coriandrum sativum* L.) fruits. *Scientia Horticulturae*, 122, 116–124.

Myers RH, Montgomery DC, Anderson-Cook CM (2009). *Response Surface Methodology: Process and Product Optimization Using Designed Experiments* (3rd ed.). Chichester, UK: Wiley.

Olden JD, Jackson DA (2002). Illuminating the "black box": a randomization approach for understanding variable contributions in artificial neural networks. *Ecological Modelling*, 154, 135–150.

Paixao Coelho JA, Figueiredo Palavra AM (2015). Supercritical fluid extraction of compounds from spices and herbs. In T Fornari, R. P. Stateva (Eds.), *High pressure fluid technology for green food processing* (pp. 357–396). New York: Springer.

Pandey A, Bigoniya P, Raj V, Patel KK (2011). Pharmacological screening of *Coriandrum sativum* Linn. for hepatoprotective activity. *Journal of Pharmacy & Bioallied Sciences*, 3(3), 435–441.

Paniwnyk L, Beaufoy E, Lorimer JP, Mason TJ (2001). The extraction of rutin from flower buds of *Sophora japonica*. *Ultrasonics Sonochemistry*, 8, 299–301.

Pavlić B, Vidović S, Vladić J, Radosavljević R, Zeković Z (2015). Isolation of coriander (*Coriandrum sativum* L.) essential oil by green extractions versus traditional techniques. *The Journal of Supercritical Fluids*, 99, 23–28.

Pourmortazavi SM, Hajimirsadeghi SS (2007). Supercritical fluid extraction in plant essential and volatile oil analysis. *Journal of Chromatography A*, 1163, 2–24.

Radojković M, Zeković Z, Mašković P, Vidović S, Mandić A, Mišan A, Đurović S (2016). Biological activities and chemical composition of *Morus* leaves extracts obtained by maceration and supercritical fluid extraction. *The Journal of Supercritical Fluids*, 117, 50–58.

Ramadan MF, Kroh LW, Mörsel JT (2003). Radical Scavenging Activity of Black Cumin (*Nigella sativa* L.), Coriander (*Coriandrum sativum* L.), and Niger (*Guizotia abyssinica* Cass.) Crude Seed Oils and Oil Fractions. *Journal of Agricultural and Food Chemistry*, 51, 6961–6969.

Ramić M, Vidović S, Zeković Z, Vladić J, Cvejin A, Pavlić B (2015). Modeling and optimization of ultrasound-assisted extraction of polyphenolic compounds from Aronia melanocarpa by-products from filter-tea factory. *Ultrasonics Sonochemistry*, 23, 360–368.

Ravi R, Prakash M, Bhat KK (2007). Aroma characterization of coriander (*Coriandrum sativum* L.) oil samples. *European Food Research and Technology*, 225, 367–374.

Rostagno MA, Palma M, Barroso CG (2003). Ultrasound-assisted extraction of soy isoflavones. *Journal of Chromatography A*, 1012, 119–128.

Sahib NG, Anwar F, Gilani AH, Hamid, AA, Saari, N, Alkharfy KM (2012). Coriander (*Coriandrum sativum* L.): A Potential Source of High-Value Components for Functional Foods and Nutraceuticals-A Review. *Phytotherapy Research*, 27, 1439-1456.

Shahwar MK, El-Ghorab AH, Anjum FM, Butt MS, Hussain S, Nadeem M (2012). Characterization of Coriander (*Coriandrum sativum* L.) Seeds and Leaves: Volatile and Non Volatile Extracts. *International Journal of Food Properties*, 15, 736–747.

Shokri A, Hatami T, Khamforoush M (2011). Near critical carbon dioxide extraction of Anise (*Pimpinella Anisum* L.) seed: Mathematical and artificial neural network modeling. *The Journal of Supercritical Fluids*, 58, 49–57.

Singh G, Maurya S, de Lampasona MP, Catalan CAN (2006). Studies on essential oils, Part 41. Chemical composition, antifungal, antioxidant and sprout suppressant activities of coriander (*Coriandrum sativum*) essential oil and its oleoresin. *Flavour and Fragrance Journal*, 21, 472–479.

Sodeifian G, Sajadian SA, Saadati Ardestani N (2016). Optimization of essential oil extraction from *Launaea acanthodes* Boiss: Utilization of supercritical carbon dioxide and cosolvent. *The Journal of Supercritical Fluids*, 116, 46–56.

Sourmaghi MHS, Kiaee G, Golfakhrabadi F, Jamalifar H, Khanavi M (2015). Comparison of essential oil composition and antimicrobial activity of *Coriandrum sativum* L. extracted by hydrodistillation and microwave-assisted hydrodistillation. *Journal of Food Science and Technology*, 52, 2452–2457.

Sriti J, Talou T, Wannes WA, Cerny M, Marzouk B (2009). Essential oil, fatty acid and sterol composition of Tunisian coriander fruit different parts. *Journal of the Science of Food and Agriculture*, 89, 1659–1664.

Sriti J, Wannes WA, Talou T, Vilarem G, Marzouk B (2011). Chemical Composition and Antioxidant Activities of Tunisian and Canadian (*Coriandrum sativum* L.) Fruit. *Journal of Essential Oil Research*, 23, 7–15.

Tchaban T, Taylor MJ, Griffin JP (1998). Establishing impacts of the inputs in a feedforward neural network. *Neural Computing & Applications*, 7, 309–317.

Teixeira B, Marques A, Ramos C, Neng NR, Nogueira JMF, Saraiva JA, Nunes ML (2013). Chemical composition and antibacterial and antioxidant properties of commercial essential oils. *Industrial Crops and Products*, 43, 587–595.

Tufeu R, Subra P, Plateaux C (1993). Contribution to the experimental determination of the phase diagrams of some (carbon dioxide+a terpene) mixtures. *The Journal of Chemical Thermodynamics*, 25, 1219–1228.

Veličković V, Đurović S, Radojković M, Cvetanović A, Švarc-Gajić J, Vujić J, ... Mašković PZ (2017). Application of conventional and non-conventional extraction approaches for extraction of *Erica carnea* L.: chemical profile and biological activity of obtained extracts. *The Journal of Supercritical Fluids*, 128, 331-337.

Wang L, Weller CL (2006). Recent advances in extraction of nutraceuticals from plants. *Trends in Food Science & Technology*, 17, 300–312.

Wangensteen H, Samuelsen AB, Malterud KE (2004). Antioxidant activity in extracts from coriander. *Food Chemistry*, 88, 293–297.

Wong P, Kitts D. (2006). Studies on the dual antioxidant and antibacterial properties of parsley (*Petroselinum crispum*) and cilantro (*Coriandrum sativum*) extracts. *Food Chemistry*, 97, 505–515.

Yoon Y, Swales G, Margavio TM (1993). A Comparison of Discriminant Analysis versus Artificial Neural Networks. *Journal of the Operational Research Society*, 44, 51–60.

Zahedi G, Azarpour A (2011). Optimization of supercritical carbon dioxide extraction of Passiflora seed oil. *The Journal of Supercritical Fluids*, 58, 40–48.

Zeković Z, Bera O, Đurović S, Pavlić B. (2017). Supercritical fluid extraction of coriander seeds: Kinetics modelling and ANN optimization. *The Journal of Supercritical Fluids*, 125, 88–95.

Zeković Z, Bušić A, Komes D, Vladić J, Adamović D, Pavlić B (2015). Coriander seeds processing: Sequential extraction of non-polar and polar fractions using supercritical carbon dioxide extraction and ultrasound-assisted extraction. *Food and Bioproducts Processing*, 95, 218–227.

Zeković Z, Cvetanović A, Švarc-Gajić J, Gorjanović S, Sužnjević D, Mašković P, Savić S, Radojković M, Đurović S (2017). Chemical and biological screening of stinging nettle leaves extracts obtained by modern extraction techniques. *Industrial Crops and Products*, 108, 423–430.

Zeković Z, Đurovic S, Pavlić B (2016). Optimization of ultrasound-assisted extraction of polyphenolic compounds from coriander seeds using response surface methodology. *Acta Periodica Technologica*, 47, 249–263.

Zeković Z, Pavlić B, Cvetanović A, Đurović S (2016). Supercritical fluid extraction of coriander seeds: Process optimization, chemical profile and antioxidant activity of lipid extracts. *Industrial Crops and Products*, 94, 353–362.

Zeković Z, Vidović S, Vladić J, Radosavljević R, Cvejin A, Elgndi MA, Pavlić B (2014). Optimization of subcritical water extraction of antioxidants from *Coriandrum sativum* seeds by response surface methodology. *The Journal of Supercritical Fluids*, 95, 560–566.

Zeković Z, Vladić J, Vidović S, Adamović D, Pavlić B (2016). Optimization of microwave-assisted extraction (MAE) of coriander phenolic antioxidants - response surface methodology approach. *Journal of the Science of Food and Agriculture*, 96, 4613–4622.

Zoubiri S, Baaliouamer A (2010). Essential oil composition of *Coriandrum sativum* seed cultivated in Algeria as food grains protectant. *Food Chemistry*, 122, 1226–1228.

In: Coriander
Editor: Deepak Kumar Semwal

ISBN: 978-1-53616-483-1
© 2019 Nova Science Publishers, Inc.

Chapter 6

POLYPHENOLIC COMPOUNDS OF CORIANDER PLANT FOR HUMAN HEALTH AND DISEASES

Arjun Pandian[1],*, Raju Ramasubbu[2], Kaliyaperumal Ashokkumar[3], Ruchi Badoni Semwal[4], Sudharshan Sekar[5] and Samiraj Ramesh[6]

[1]Department of Biotechnology, PRIST Deemed University,
Vallam, Thanjavur, Tamil Nadu, India
[2]Department of Biology, The Gandhigram Rural Institute
(Deemed to be University), Gandhigram, Dindigul, Tamil Nadu, India
[3]Cardamom Research Station, Kerala Agricultural University,
Pampadumpara, Idukki, Kerala, India
[4]Department of Chemistry, Pt. Lalit Mohan Sharma Government
Post-Graduate College, Rishikesh, Uttarakhand, India
[5]Department of Biotechnology and Food Technology,
University of Johannesburg, Johannesburg, South Africa
[6]Department of Microbiology, PRIST Deemed University, Vallam,
Thanjavur, Tamil Nadu, India

*Corresponding Author's Email: arjungri@gmail.com.

Abstract

Coriander (*Coriandrum sativum L.*) is an annual, culinary, aromatic and medicinal plant. The leaves and seeds of this plant are used in food, pharmaceutical and cosmetic industries. The medicinal properties of the seeds are mainly digestion, rheumatism, joint pains and against worms. It has an array of pharmacological effects such as anticancer, anti-hyperglycemic, antifertility, anti-inflammatory, anxiolytic, antispasmodic, antihyperlipidemic, digestive stimulant and hypotensive. Its fruits are considered to be antibilious, diuretic, carminative, refrigerant, stomachic and aphrodisiac. The plant contains a variety of secondary metabolites including polyphenols such as dimethoxycinnamoylhexoside, quercetin-3-O-rutinoside, quercetin 3-O-glucuronide, kaempferol-3-O-rutinoside and quercetin-3-O-glucoside. Many important flavonol derivatives like 3-O-caffeoylquinic acid, caffeoylquinic acid, ferulic acid glucoside and p-coumaroylquinic acid have also been reported from this plant. This chapter reports a comprehensive knowledge of the coriander plant including the traditional uses, pharmacology and chemistry.

Keywords: antifertility, anticancer, kaempferol, proanthocyanidins, quercetin

Introduction

Traditional herbal medicine has been adopted as a main course of treatment since time immemorial and still acceptable throughout the world. Due to its unique applications, the researchers have paid attention to work on this area of medical sciences to discover new bioactives. Consequently, abundant studies have been regulated on a number of medicinally important plants and have been paying attention to the bioactive compounds and their biological activities (Properzi et al., 2012).

An illustration of an important medicinal plant coriander (*Coriandrum sativum*), belongs to the family Apiaceae, showed that it is an annual herb and a native of the Eastern Mediterranean region. The plant is now extensively cultivated in several other parts of the world including Russia, Asian countries and Central Europe (Sahib et al., 2013). It is an extensively disseminated and mainly cultivated for its seeds which are the popular spice of the kitchen. It is reported to have numerous traditional uses and displays various pharmacological effects like diarrhoea, vomiting, cough, fever, dysentery and assorted inflammatory conditions as confirmed by various studies (Sahib et al., 2013).

Coriander green fresh leaf is commonly recognized as Dhania, Kothamalli, Chinese aromatic plant or Cilantro. It is extensively featured in cuisines of different countries including India, Mexico and China. Due to the presence of essential oil (EO), coriander leaves acquire a distinctive aroma and also used as a food flavouring agent (Gil et al., 2002). The leaves of the plant are essentially used as a source for the EO. The dried seeds are added in dishes, fresh leaves used in South India to prepare rasam as a pungent spice, and it's considered as a very good agent for digestion. The leaves are also a significant component in the Thai and Vietnamese cuisine (Gil et al., 2002). In India, the dry seeds are the most important ingredient of the curry powder. In addition, seeds are used for flavouring of numerous foods like fish, meat, bakery and also confectionery product (Gil et al., 2002).

TRADITIONAL MEDICINAL USES

It has been extensively used as cooking constituent and a traditional remedy for diverse disorders. All parts of this herb are edible, extremely dissimilar in use and flavour (Bhat et al., 2014). The coriander root has a different flavour than that of the leaf and frequently used in Asian cuisines, while the chopped stems are used in the form of soups and stews (Verma et al., 2011). Owing to the pungent flavour and health benefits, it has confirmed its worth as an imperative medicinal and aromatic plant as reported by different herbologists. It's traditionally used for smallpox, gastric complaints, anaemia, nausea, fever, cold, measles and hernias. Seeds are used treating frequent digestion complaints; nausea, dyspepsia, and also dysentery. Leaves are useful in improving digestion. The leaves contain minerals like Fe, Mg and Mn, and also rich in vitamins A, B, and C. It's a good source of dietary fibre. In Indian traditional medicine, it's used in different types of disorders like urinary and respiratory problems. In addition, coriander is used as a diuretic, diaphoretic and carminative agent. In Turkey, the seeds are used as a digestive, appetizer and carminative agent (Ugulu et al., 2009).

According to Ayurveda, the regular use of seeds decoction is useful for lowering the hyperlipidemia, cholesterol and triglycerides levels in the blood (Lal et al., 2004). It is also used for joint pains and inflammation. Based on the traditional knowledge, use of Maharasnadhi Quather, an absolute conventional polyherbal formulation contains coriander seeds as a major component, is suggested to be an effective Ayurvedic remedy for joint pains in arthritic conditions (Thabrew et al., 2003).

SECONDARY METABOLITES IN SEEDS

The seeds of the coriander plant contain different bioactive compounds like fatty acids, tocols, sterols, and essential volatile compound.

Lipids

The seeds containing the highest amount of fatty acids are petroselinic (80%), linoleic (16%), oleic (7%), palmitic (4%) and stearic acid (3%) whereas the minor fatty acids are almitoleic, α-linolenic, arachidic, gadoleic, erucic and docosahexenoic acid (Msaada et al., 2009). The neutral lipids in seed oil have been characterised largely as triacylglycerols (95%) followed by free fatty acids (2%), diacylglycerols (1%) and diacylglycerols (0.5%) (Sriti et al., 2010). The study by Sriti et al. (2010) also suggested that the polar lipids present in the seeds are phospholipid (36%), phosphatidyl ethanolamine (34%), phosphatidylinositol (15%), phosphatidic acid and phosphatidylglycerol as the least. Moreover, digalactosyldiacylglycerol (62%) and monogalactosyldiacylglycerol (37%) were also reported from the seeds.

Sterols and Tocols

The coriander seed oil represents imperative resources for sterols. It contains stigmasterol (21.7–29.8%) and β-sitosterol (24.8–36.8%) as major contents. In addition, γ-tocopherol, δ-tocopherol and α-tocopherol were also reported from the oil (Sriti et al., 2010).

Essential oil Compounds

Essential oil (EO) extraction of coriander seeds is mainly performed through hydrodistillation or steam distillation process. EO yield of Indian coriander was recorded ranging from 0.18% to 0.39%. Whereas, Tunisian coriander was found to 0.35% (Msaada et al., 2007). It majorly contains linalool (up to 88%) followed by nerol, borneol, geraniol, α-terpinene, myrcene, α-pinene, linalool acetate and β-pinene (Abou El-Nasr et al., 2013). The major components reported from the EO of the fresh coriander leaves were trans 2-dodecenal, 2-methylenecyclopentanol, decanal, dodecanal, 2-tridecenoic acid, cyclooctane and 2-octenal (Arjun et al., 2017) (Figure 1).

Polyphenols

Msaada et al. (2014) reported that Syrian coriander contains maximum total phenolic content (1.09 mg GAE/g, DW) followed by Tunisian (1.00 mg) and Egyptian coriander (0.94 mg). An ethyl acetate seeds extract of Norwegian coriander was found to contain 1.89 GAE/100g total phenolic content (Wangensteen et al., 2004). The flavonoid and condensed tannins estimated in the methanolic seeds extract ranged from 2.03 to 2.51 mg CE/g DW. In the phenolic fraction, the major phenolic acids found were gallic, vanillic, chlorogenic, caffeic, p-coumaric, ferulic, rosmarinic, trans-hydroxycinnamic, O-coumaric, salicylic and trans-cinnamic acids.

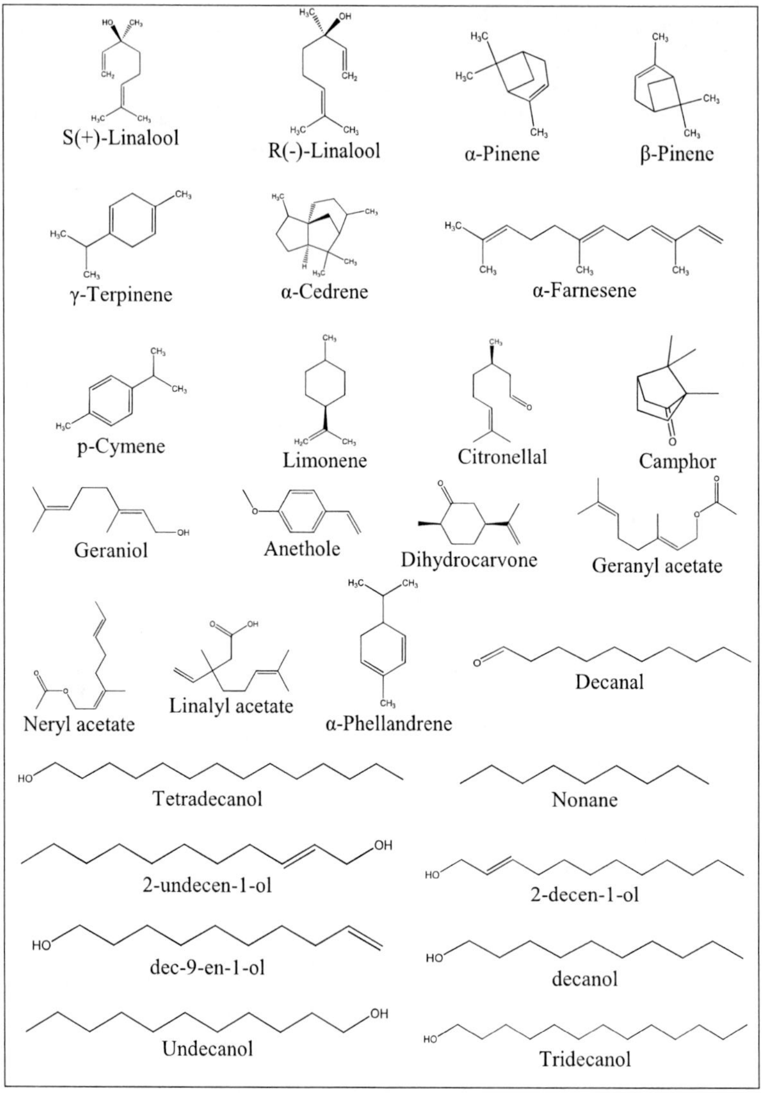

Figure 1. EO compounds present in *Corianderum sativum*.

Moreover, the major flavonoids were quercetin-3-rhamnoside, quercetin dihydrate, luteolin, rutintrihydrate, kaempferol, resorcinol, apigenin, naringin, coumarin and flavone (Msaada et al., 2014). Coriander leaves are mainly containing volatile oil together with phenolic compounds including phenolic acids, flavonoids and polyphenols (Matasyoh et al., 2009).

SECONDARY METABOLITES IN LEAVES

Coriander leaves are mainly containing volatile oil together with phenolic compounds including phenolic acids, flavonoids and polyphenols (Matasyoh et al., 2009).

Lipids

Total fatty acid contents in the basal and upper side of leaves were estimated and found that basal side contains 61.21 mg/g DW whereas upper side contains 41.8 mg/g DW (Neffati and Marzouk, 2008). It contains a predominance of polyunsaturated fatty acids like α-linolenic, heptadecenoic, linoleic, and palmitic acids whereas stearic, stearidonic, oleic, cis- and trans-palmitoleic acids were found in trace amount. The ether extract of the leaves showed the presence of β-cryptoxanthin epoxide, lutein-5,6-epoxide, β-carotene, neoxanthin and violaxanthin. It was also found a good source of β-carotene (Divya et al., 2012).

Polyphenols

Polyphenolic acids in coriander leaf include dimethoxy-cinnamoyl-hexoside, quercetin-3-O-rutinoside, kaempferol-3-O-rutinoside, quercetin 3-O-glucuronide, quercetin-3-O-glucoside. Four flavonol derivatives include 3-O-caffeoylquinic, caffeoylquinic acids, ferulic acid glucoside, and *p*-coumaroylquinic acids. In coriander, vegetative parts derivatives of quercetin were chief bioactive compounds originated. In aerial parts, 21 phenolic compounds are identified. Among those, the main are coumarins, flavonoids, and phenolcarboxylic acids. Other compounds include luteolin, apigenin, hyperoside, vicenin, hesperidin, diosmin, dihydroquercetin, catechin, orientine, chrysoeriol, gallic, salicylic, ferulic acid, dicoumarin, 4-hydroxycoumarin, esculin, esculetin, tartaric acid, maleic acid and arbutin identified (Oganesyan et al., 2007).

The fresh coriander leaves obtained from India contains 24.02 mg GAE/g total phenol contents (Arjun et al., 2017). The major polyphenolic acids identified in leaves of Indian coriander are *p*-coumaric, vanillic, cis- and trans-ferulic acids. In addition, flavonoids in a leaf comprise kaempferol, quercetin, 3'-OMe quercetin, acacetin and also 4'-OMe quercetin. Glycoflavones has not been detected. Therefore, Indian coriander leaf has a high-quality source of quercetin (Nambiar et al., 2010). Melo et al. (2005) reported that the phenolic acids in the coriander leaves from Brazil have protocatechinic acid (6.43 µg/mL), caffeic acid (4.34 µg/mL), and glycitin (3.27 µg/mL). It contains microelements, particularly Zn, and anthocyanins which improves biosynthesis through salicylic acid. In dissimilarity, N appreciably decreased anthocyanin pleased to it is in lowly velocity. Also, P and K, a negative consequence of anthocyanin contented and decreased (Rahimi et al., 2013). The in vitro culture used to investigate novel industrial, medicinal and pharmaceutical potentialities, such as secondary metabolites production are flavones, flavonols and anthocyanins (Rahimi et al., 2013).

Essential Oil Compounds

The seeds of the plant yields comparatively higher essential oil than that of leaves and roots. Among the 44 chemical components of leaf EO, mostly are aromatic acid in which 2-decenoic (30.8%), E-11-tetradecenoic (13.4%), capric acids (12.7%) are the major ones (Bhuiyan et al., 2009). Kenyan coriander leaf contains aldehydes and alcohol as 56.1% and 46.3%, respectively. The major constituents of this oil included (E)-2-decenal (15.9%), decanal (14.3%), (E)-2-decen-1-ol (14.2%) and n-decanol (13.6%), while supplementary compounds included (E)-2-tridecen-1-al, undecanol, (E)-2-dodecenal, dodecanal and undecanal (Matasyoh et al., 2009). The main compounds in Brazilian coriander were reported to 1-decanol (24.20%), (E)-2-decenol (18.00%), and (Z)-2-dodecenol (17.60%) together with aldehydes (Begnami et al., 2010). On the other hand, the main volatile components of Indian coriander leaf were recorded to (E)-2-decenal (18%), decanal (14%), dec-9-en-1-ol (11%), (E)-2-dodecenal (8%), n-tetradecanol (6%), dodecanal (5%) and decanol (5%). The Korean coriander leaves revealed 39 components represented by 99.62% of the total oil. Its major components were found as cyclododecanol (23%), tetradecanal (17%), 2-dodecenal (9%), 1-decanol (7%), 13-tetradecenal (6%), 1-dodecanol (6%), dodecanal (5%), 1-undecanol (2%), and decanal (2%) (Padalia et al., 2011).

BIOLOGICAL ACTIVITIES

The coriander secondary metabolites in extracts and EO showed different biological activities such as antioxidant, antimicrobial, antihypertensive, antimutagenic, antidiabetic and diuretic (Matasyoh et al., 2009). Volatile components in EO from both leaves and seeds reported to inhibit the growth of different microorganisms. Apart from the medicinal and pharmacological properties, coriander also reported having adverse effects such as convulsion, appetite suppression, anxiety, insomnia and dyspeptic complaints (Eidi et al., 2009). Earlier phytochemical analysis on different parts revealed largely EO (Emamghoreishi et al., 2005) together with terpenoid glycosides, polyphenols, coumarins, and fatty acids (Arjun et al., 2017).

Analgesic Activity

The aqueous extract of coriander seeds inhibited the central pain receptors revealing its analgesic activity (Pathan et al., 2011). Major compounds in coriander namely linalool and monoterpene alcohol play an important role like analgesic activity. Experiments on mice suggest that the glutamatergic arrangement in antinociception was contributed by linalool in antinociception elicited through linalool (Batista et al., 2008). β-cyclodextrin complexes in linalool confirm the production of antinociceptive consequence greater to linalool in (Quintans et al., 2013).

Antimicrobial Activity

Coriander EO shows a broad spectrum inhibition of antibacterial and antifungal agent. Antibacterial activity of seed extract was seen alongside gram-positive and gram-negative bacteria like *Staphylococcus aureus* (PTCC1431), *Pseudomonas aeruginosa* and *Klebsiella pneumonia*. "Plantaricin CS", a narrative antimicrobial peptide having a broad range of antibacterial action was inaccessible from leaf extract against on *S. aureus* (MIC = 1.3 mg/mL). The germicidal property was also seen against *K. pneumoniae* and *P. aeruginosa* with MIC = 2.65 mg/mL and 3.2 mg/mL respectively. The polysaccharides in the cell walls of bacteria prevents active compounds accomplishment to the cytoplasmic membrane, therefore, less antimicrobial activity of Plantaricin CS was seen against gram-negative bacteria Action moderately similar to the activity of antibiotics such as ofloxacin, gentamicin sulfate, tobramycin and beneath analogous conditions ciprofloxacin screened, leaf EO exhibited outstanding activity alongside gram-positive and gram-negative bacteria (Zare-Shehneh et al., 2014). Antibacterial activity potential of EO alongside gram-positive and gram-negative bacteria is based on its membrane permeability. The leaf EO, chiefly owing to the extended chain of C (C6–C10) alcohols and aldehydes, is efficient against *Listeria monocytogenes* (Joji Reddy et al., 2012).

The essential oil of leaves showed antifungal activity against *Fusarium oxysporum, Curvularia pallescens, F. moniliforme, Aspergillus niger, A. terreus*, and *F. graminearum*. Plantaricin CS antifungal activity against *Penicillium lilacinum* and *A. niger* with MIC values of 2.5 mg/mL and 2.3 mg/mL, respectively (Zare-Shehneh et al., 2014). Linalool inhibited the growth of *Candida* and *Trichophyton* with MIC ranging from 0.03 to 2 mg/mL. Synergistic effect was observed in the combination of linalool and ketoconazole (FICs = 0.06 - 0.53 mg/mL). EO prevents infection by Candida yeast infection (Furletti et al., 2011).

Anxiolytic Activity

The seed extract showed anxiolytic and relaxant effects. Additional chemical and pharmacological investigations are mandatory to clarify the accurate mechanistic approach of seed extracts and isolate its energetic principles. The anxiolytic commotion seems feasible to be connected with its EO content and flavonoids (Mahendra and Bisht, 2011).

Anthelmintic Activity

In vivo and *in vitro* assessment showed the anthelmintic activity of coriander seed extract against a nematode parasite *Haemonchus contortus*. The hydro-alcoholic extract showed improved in vitro action by touching mature parasites as compared to aqueous solitary (Eguale et al., 2007).

Hypoglycemic and Hypolipidemic Activities

Efficacy of coriander extracts against diabetes has been verified in previous studies (Waheed et al., 2006). The hypoglycemic activity of EO may attribute synergistic accomplishment of geranyl acetate, linalool and γ-terpinene (Abou El-Soud et al., 2012). Anti-hyperglycemic mechanisms of coriander are associated through insulin secretion stimulus and glucose uptake enhancement. Coriander is considered as an impending foundation of functional nutritional supplements for humanizing, controlling blood glucose and preventing the symptoms of chronic complications in type II diabetes mellitus (Pandeya et al., 2013). Tahraoui et al. (2007) reported that the seeds and leaves are used as an antidiabetic representing the mechanism for controlling hyperglycemia. The hypoglycemic effect of coriander leaves was reported through 20 diabetes-induced rats. Out of the 4 groups of rats, three groups were supplemented with concerning 15 g (60 g/kg BW/d) leaves for 15 days. The fourth diabetic untreated group (positive control) and a non-diabetic group (negative control) had a conventional standard cut down. The experimental results showed that the leaf consumption did not produce a significant hypoglycemic effect in diabetic rats (Jelodar et al., 2007).

Coriander seeds have hypolipidemic activity through a diverse aspect of lipid metabolism in the experimental animals. It degrades bile acids and unbiased sterols, thus lowering cholesterol in tissues and serum (Dhanapakiam et al., 2008). Bioactive compounds present in the seeds are the cause of the hypolipidemic activity in the seeds. Fatty acids like linoleic, palmitic, oleic, stearic and ascorbic acid reduce not only the cholesterol level in blood but also the cholesterol deposition in the internal walls of veins and arteries (Ertas et al., 2005). Coriander is a potential popularized household herbal medicine having defensive and healing consequence alongside hyperlipidemia (Lal et al., 2004).

EO of coriander seeds plays a significant function in stored grain fortification and decreases risks allied with the use of the synthetic insecticide. It develops Canister into an attractive to conserve chemical control strategy (Khani and Rahdari et al., 2012). Insecticidal movement with rice pests (*Rhyzopertha dominica, Cryptolestes pusillus* and *Sitophilus oryzae*) were tested in lab condition for their unpredictable toxicity. Linalool is the chief energetic component of EO containing 1617 ppm alongside three pests. Affluent fractions of camphor over 400 ppm were exceptionally toxic to *C. pusillus* and *R. dominica* (López et al., 2008). The seed oil has remarkable toxic effects alongside the *Aedesaegypti larvae* LC_{50}; 21.5 ppm and significantly might function as immunotoxicity alongside the insects (Chung et al., 2012). The seed extract and EO of coriander acquire a tranquillizer hypnotic activity. The main energetic components of coriander in water extract are accountable for a hypnotic consequence (Emamghoreishi and Heidari-Hamedani, 2006).

Antioxidant Activity

Antioxidants refer to a collection of compounds that are able to a setback or inhibit the lipids oxidation, biomolecules and consequently stop or repair the smashed up human body cells (Shahidi and Naczk, 2004; Tachakittirungrod et al., 2007).

According to free radical scavenging biology and medical science, the reactive oxygen species (ROS) are contributory agents for numerous physical conditions. Standard ingestion of antioxidants can treat such physical condition issues and decrease ROS production in the human body (Tachibana et al., 2001; Arjun et al., 2017). In recent years, consumers are concerned about the addition of synthetic additives to the food, antioxidants, butylated hydroxyl anisole (BHA) and butylated hydroxyl toluene (BHT) that induces DNA damage. Interestingly, medicinal plants hold abundant bioactive compounds having a potential anti-oxidative activity that reduces oxidative stress-induced wounds (Arjun et al., 2017). Herb and spices; have several phytochemicals and are a resource of natural antioxidants such as flavonoids, phenolic compounds, phenolic diterpenes, tannins, alkaloids and phenolic acids. The expected antioxidants are recognized to guard cells against oxidative stress-induced injury, which is normally well-thought-out to be a reason for ageing, cancer and degenerative diseases (Ringman et al., 2005). These physical conditions promote possessions of antioxidants from the plants and spices along with their defensive effect that counteracts with the ROS. Coriander shoot fraction contains caffeic acid; a phenolic correlated to the derivatives of hydroxyl cinnamic acids groups. It is the most important component with the antioxidant activity (Godow et al., 1997). Quinic acid is measured as a phenolic precursor of numerous aromatic compounds in the metabolism of vegetables. This consequence, according to Zhang et al. (2001), was less than that of chlorogenic acid, rutin and quercetin, however analogous to caffeic acid (Masella et al., 1999) and (Harborne, 1973). Ortho dihydroxy benzene arrangements facilitate the contribution of H and continuation of an unsaturated aliphatic sequence located on the aromatic ring, which increases its constancy of the phenoxy free radical all the way through reverberation (Lu and Foo, 2001). Protocatechinic acid inhibited the human low-density lipoprotein (LDL) catalyzed through copper (Zhang, et al., 2001). Xanthophylls are β-cryptoxanthin epoxide, lutein 5, 6-epoxide, violaxanthin and neoxanthin,

(Guerra et al., 2005). β-cryptoxanthin epoxide, β-carotene, lutein 5, 6 epoxide (Rodriguez-Amaya, 1999b).

Anticancer Activity

EO of the seed, especially linalool is a major component showing anticancer activity. As expected the compound inhibits cell proliferation reasonably and develops a therapeutic directory of anthracyclines in the administration of human breast cancer, particularly in multidrug challenging tumours (Ravizza et al., 2008). Coriander leaves and seeds extracts showed *in vitro* antitumor and immune-modulating activity whereas root extracts showed anti-proliferative activity against human breast cancer cell lines (Table 1). This put forward its potential in the prevention of cancer and metastasis inhibition (Gomez-Flores et al., 2010). The anti-tumorigenic properties of coriander accredit its defensive function alongside the injurious property in the metabolism of lipids connected through this melanoma in investigational colon cancer cells. The functional supplements used in combination with conservative drugs improve the treatment of cancer (Chithra and Leelamma, 2000).

Activity against Indigestion

The coriander has been justified for the comprehensive animal study. The digestive refreshment action might be measured through liver stimulation to secrete more bile enriched in bile acids, and enzyme encouragement activities contributing indigestion, in cooperation of intestinal and pancreatic origin. Such motivation activities of digestive enzymes lead to an accelerated in general digestive process (Platel and Srinivasan, 2004).

Table 1. Pharmacological Studies for Coriander sativum

S.No.	Activity	Extract type	Models/dosage used	References
1	Antibacterial	Fruits EO	In vitro, bacterial skin infections (0.04–025% v/v)	Casetti et al., 2012
2	Antifungal	Leaves EO	In vitro and micro-dilution techniques (1000–0.48 µg/mL)	Freires et al., 2014
3	Antioxidant	EtOH, MeOH, DCM, extracts of seeds & leaves	In vitro DPPH assay, inhibition of 15-lipoxygenase, inhibition of Fe^{2+} induced porcine brain, phospholipid peroxidation (167 µg/mL)	Al-Mofleh et al., 2006
4	Anticancer	MeOH and H_2O extracts of leaves and seeds	In vitro, mouse, lymphoma cell line L5178Y-R (7.8–125 µg/mL)	Gomez-Flores et al., 2010
5	Anthelmintic	H_2O and H_2O-EtOH extracts of seeds	In vitro, in vivo study in infected sheep with Haemonchus contortus (0.5–0.12–0.18 mg/mL, 0.45 and 0.9 g/kg)	Eguale et al., 2007
6	Anti-convulsant	H_2O and EtOH extracts of seeds	In vivo in mice, pentylenetetrazole and maximal electroshock tests (H_2O Ex. 0.05, 0.2, 0.35 & 0.5 g/kg; EtOH: 0.5, 2.0, 3.5 & 5 g/kg)	Chithra and Leelamma, 2000
7	Anti-inflammatory	Aerial parts EO	Clinical study, ultraviolet erythema test (0.5% and 1.0%)	Reuter et al., 2008
8	Anxiolytic	H_2O extract of seeds	In vivo in mice, elevated plus-maze test (50, 100 and 200 mg/kg)	Pathan et al., 2011
9	Hypo-lipidemic	H_2O extract of seeds	In vivo in rats, normal and obese-hyperlipidemic; hypercaloric diet and forced limited physical activity in rats (20 mg/kg, daily dosing for 30 d, sub-chronic study)	Aissaoui et al., 2011
10	Insecticidal	Aerial parts EO	In vitro study of eggs, larvae and adults of Tribolium Castaneum. Filter paper arena test (2, 4, 8 and 12 µg dissolved in 1 mL acetone)	Islam et al., 2009
11	Memory-enhancing	Fresh leaves	In vivo in rats, elevated plus-maze served as exteroceptive behavioural model (5, 10 and 15% w/w in normal animal diet 45 d)	Mani and Parle, 2009
12	Sedative hypnotic	Seeds EO, H_2O and H_2O-EtOH extracts	In vivo in mice, treatments were carried out for 30 min (100, 200, 400 and 600 mg/kg)	Emamghoreishi and Heidari-Hamedani, 2006

Conclusion

The current assessment summarizes a number of frequent reports on the phytochemical composition of seeds and aerial parts (herb) along with their different biological activity. Bioactive constituent stating modern journalism and supports coriander potentially as an important medicinal plant. It has unfolded its use as conventional medicine that has been engaged medicinally in unthinking airway diseases such as bronchiolitis and asthma. The broad spectrum of pharmacological effects *in vivo* and *in vitro* studies have been performed on this medicinal plant. It has numerous biological properties like antioxidant, antimicrobial, hypoglycemic, anxiolytic, hypolipidemic, analgesic, anticonvulsant, anti-inflammatory and anticancer activities.

In addition, a synergic consequence has been established for the antibacterial and antifungal activity of EO and conventional antibiotics/antifungal agents show the potential antifungal activity of the coriander EO as a potential source for the management of oral diseases.

Moreover, as a supplement, Dhania might be functional in amalgamation with predictable drugs to improve the treatment of cancer and Alzheimer disease. Optimistically, the bioactive constituents of Dhania seeds and leaves and their different types of biological activities determined are accommodating to generate more attention towards coriander through important novel clinical and pharmacological applications and therefore, might be functional in mounting novel drug formulations in opportunity.

References

Abou El-Nasr T. H. S., Ibrahim, M. M., Aboud, K. A., and Magda El-Enany, A. M. 2013. Assessment of Genetic Variability for Three Coriander (*Coriandrum sativum* L.) Cultivars Grown in Egypt, Using morphological Characters, Essential Oil Composition and ISSR Markers. *World Appl Sci J*, 25 (6): 839-849.

Abou El-Soud, N. H., El-Lithy, N. A., El-Saeed, G. S. M., Wahby, M. S., Khalil, M. Y., Abou El-Kassem, L. T. 2012. Efficacy of *Coriandrum sativum* L. essential oil as antidiabetic. *J Appl Sci Res*, 8(7): 3646–55.

Aissaoui, A., Zizi, S., Israili, Z. H., and Lyoussi, B. 2011. Hypoglycemic and hypolipidemic effects of *Coriandrum sativum* L. In Merionesshawi rats. *J Ethnopharmacol*, 137(1):652–61.

Al-Mofleh, I. A., Alhaider, A. A., Mossa, J. S., Al-Sohaibani, M. O., Rafatullah, S., and Qureshi, S. 2006. Protection of gastric mucosal damage by *Coriandrumsativum* L. pretreatment in Wistar albino rats. *Environ Toxicol Pharmacol*, 22(1):64–9.

Arjun, P., Semwal, D. K., Semwal, R. B., Malaisamy, M., Sivaraj, C., and Vijayakumar, S. 2017. Total Phenolic Content, Volatile Constituents and Antioxidative Effect of *Coriandrum sativum, Murraya koenigii* and *Mentha arvensis*. *Nat Prod J*, 7: 65-74.

Batista, P. A., Werner, M. F. P., Oliveira, E. C., Burgos, L., Pereira, P., and da Silva Brum, L. F. 2008. Evidence for the involvement of ionotropicglutamatergic receptors on the antinociceptive effect of (−)-linalool in mice. *Neurosci Lett*, 440: 299–303.

Begnami, A. F., Duarte, M. C. T., Furletti, V., and Rehder, V. L. G. 2010. Antimicrobial potential of *Coriandrum sativum* L. against different *Candida* species *in vitro*. *Food Chem*, 118: 74–7.

Bhat, S., Kaushal, P., Kaur, M., and Sharma, H. K. 2014. Coriander (*Coriandrum sativum* L.): processing, nutritional and functional aspects. *Afr J Plant Sci*, 8(1): 25–33.

Bhuiyan, N. I., Begum, J., and Sultana, M. 2009. Chemical composition of leaf and seed essential oil of *Coriandrumsativum* L. from Bangladesh. *Bangladesh J Pharmacol*, 4: 150–3.

Casetti, F., Bartelke, S., Biehler, K., Augustin, M., Schempp, C. M., and Frank U. 2012. Antimicrobial activity against bacteria with dermatological relevance and skin tolerance of the essential oil from *Coriandrum sativum* L. fruits. *Phytother Res*, 26(3): 420–4.

Chithra, V., and Leelamma, S. 2000. *Coriandrum sativum* effect on lipidmetabolismin 1,2-dimethyl hydrazine induced colon cancer. *J Ethnopharmacol*, 71: 457–63.

Chung, I. M., Ahmad, A., Kim, E. H., Kim, S. H., Jung, W. S., and Kim, J. H. 2012. Immunotoxicity activity fromthe essential oils of coriander (*Coriandrum sativum*) seeds. *Immunopharmacol Immunotoxicol*, 34(3): 499–503.

Dhanapakiam, P., Mini Joseph, J., Ramaswamy, V. K., Moorthi, M., and Senthil Kumar, A. 2008. The cholesterol lowering property of coriander seeds (*Coriandrumsativum*): mechanism of action. *J Environ Biol*, 29(1): 53–6.

Divya, P., Puthusseri, B., and Neelwarne, B. 2012. Carotenoid content, its stability during drying and the antioxidant activity of commercial coriander (*Coriandrumsativum* L.) varieties. *Food Res Int*, 45(1): 342–50.

Eguale, T., Tilahun, G., Debella, A., Feleke, A., and Makonnen, E. 2007. *In vitro* and *in vivo* anthelmintic activity of crude extracts of *Coriandrum sativum* against *Haemonchus contortus*. *J Ethnopharmacol*, 110(3): 428–33.

Eidi, M., Eidi, A., Saeidi, A., Molanaei, S., Sadeghipour, A., Bahar, M., and Bahar, K. 2009. Effect of Coriander seed (*Coriandrumsativum* L.) ethanol extract on insulin release from pancreatic beta cells in Streptozotocin induced diabetic rats. *Phytother. Res.*, 23(3): 404-6.

Emamghoreishi, M., and Heidari-Hamedani, G. 2006. Sedative-hypnotic activity of extracts and essential oil of coriander seeds. *Iran J Med Sci*, 31(1): 22–7.

Emamghoreishi, M., Khasaki, M., and Aazam, M. F. 2005. *Coriandrumsativum*: evaluation of its anxiolytic effect in the elevated plus-maze. *J Ethnopharmacol*, 96: 365-370.

Ertas, O. N., Güler, T., Çiftçi, M., Dalkiliç, B., and Yilmaz, O. 2005. The effect of a dietary supplement coriander seeds on the fatty acid composition of breast muscle in Japanese quail. *Rev Med Vet*, 156(10): 514–8.

Freires, I. d. A., Murata, R. M., Furletti, V. F., Sartoratto, A,, de Alencar, S. M., and Figueira, G. M. 2014. *Coriandrum sativum* L. (Coriander) essential oil: antifungal activity and mode of action on Candida spp., and molecular targets affected in human whole-genome expression. *PLoS One*, 9(6): e99086.

Furletti, V. F., Teixeira, I. P., Obando-Pereda, G., Mardegan, R. C., Sartoratto, A., and Figueira, G. M. 2011. Action of *Coriandrum sativum* L. essential oil upon oral *Candida albicans* biofilm formation. *Evid Based Complement Alternat Med*, 2011: 985832.

Gil, A., De La Fuente, E. B., Lenardis, A. E., Lopez Pereira, M., Suarez, S. A., Bandoni, A., Van Baren, C., Di Leo Lira, P., and Ghersa, C. M. 2002. Coriander essential oil composition from two genotypes grown in different environmental conditions. *J. Agric. Food Chem.*, 50: 2870-2877.

Godow, A. V., Joubert, E., and Hansmann, C. F. 1997. Comparison of the antioxidant activity ofaspalath in with that of other plant phenols of rooibos tea (*Aspalathus linearis*), a-tocoferol, BHT and BHA. *J. Agric. Food Chem.*, 45: 632–638.

Gomez-Flores, R., Hernández-Martínez, H., Tamez-Guerra, P., Tamez-Guerra, R., Quintanilla-Licea, R., and Monreal-Cuevas, E. 2010. Antitumor and immune modulating potential of *Coriandrum sativum*, *Piper nigrum* and *Cinnamomum zeylanicum*. *J Nat Prod*, 3: 54–63.

Guerra, N. B., Melo, E. A., and Filho, J. M. 2005. Antioxidant compounds from coriander (Coriandrumsativum L.) etheric extract. *J Food Compos Anal*, 18: 193–199.

Harborne, J. B. 1973. *Phytochemical methods*. London: Chapman & Hall, Halsted Press, a Division of John Wiley & Sons, New York.

Islam, M. S., Hasan, M. M., Xiong, W., Zhang, S. C., and Lei, C. L. 2009. Fumigant and repellent activities of essential oil from *Coriandrum sativum* (L.) (Apiaceae) against red flour beetle *Tribolium castaneum* (Herbst) (Coleoptera: Tenebrionidae). *J Pest Sci*, 82(2): 171–7.

Jelodar, G., Mohsen, M., and Shahram, S. 2007. Effect of walnut leaf, coriander and pomegranate on blood glucose and histopathology of pancreas of alloxan induced diabetic rats. *Afr J Tradit Complement Altern Med*, 4(3): 299–305.

Joji Reddy, L., Devi Jalli, R., Jose, B., and Gopu, S. 2012. Evaluation of antibacterial and DPPH radical scavenging activities of the leaf extracts and leaf essential oil of *Coriandrum sativum* Linn. *World J Pharm Res*, 1(3): 705–16.

Khani, A., and Rahdari, T. 2012. Chemical composition and insecticidal activity of essential oil from *Coriandrum sativum* seeds against *Tribolium confusum* and *Callosobruchus maculatus*. *ISRN Pharm*, 2012: 263517.

Lal, A. A., Kumar, T., Murthy, P. B., and Pillai, K. S. 2004. Hypolipidemic effect of *Coriandrumsativum* L. in triton-induced hyperlipidemic rats. *Indian J Exp Biol*, 42(9): 909–12.

López, M. D., Jordán, M. J., and Pascual-Villalobos, M. J. 2008. Toxic compounds in essential oils of coriander, caraway and basil active against stored rice pests. *J Stored Prod Res*, 44(3): 273–8.

Lu, Y., and Foo, Y. 2001. Antioxidant activities of polyphenols from sage (*Salvia officinalis*). *Food Chem*, 75: 197–202.

Mahendra, P., and Bisht, S. 2011. Anti-anxiety activity of *Coriandrum sativum* assessed using different experimental anxiety models. *Indian J Pharmacol*, 43(5): 574–7.

Mani, V., and Parle, M. 2009. Memory-enhancing activity of Coriandrumsativum in rats. *Pharmacologyonline*, 2: 827–39.

Masella, R., Cantafora, A., Modesti, D., Cardili, A., Gennaro, L., Bocca, A., and Coni, E. 1999. Antioxidant activity of 3,4-DHPEAEA and protocatechuic acid: A comparative assessment with other olive oil biophenols. *Redox Rep*, 4: 113–121.

Matasyoh, J. C., Maiyo, Z. C., Ngure, R. M., and Chepkorir, R. 2009. Chemical composition and antimicrobial activity of the essential oil of Coriandrumsativum. *Food Chem*, 113(2): 526–9.

Melo, E. A., Filho, J. M., and Guerra, N. B. 2005. Characterization of antioxidant compounds in aqueous coriander extract (*Coriandrum sativum* L.). *LWT Food Sci Technol*, 38: 15–9.

Msaada, K., Ben Jemia, M., Salem, N., Bachrouch, O., Sriti, J., Tammar, A. 2014. Antioxidant activity of methanolic extracts from three coriander (*Coriandrum sativum* L.) fruit varieties. *Arab J Chem*, 10: S3176-S3183.

Msaada, K., Hosni, K., Ben Taarit, M., Chahed, T., Kchouk, M. E., and Marzouk, B. 2007. Changes on essential oil composition of coriander (*Coriandrum sativum* L.) fruits during three stages of maturity. *Food Chem*, 102: 1131–4.

Msaada, K., Hosni, K., Ben Taarit, M., Hammami, M., and Marzouk, B. 2005. Effects of growing region and maturity stages on oil yield and fatty acid composition of coriander (*Coriandrum sativum* L.) fruit. *Sci Hortic*, 120: 525–31.

Nambiar, V. S., Daniel, M., and Guin, P. 2010. Characterization of polyphenols from coriander leaves (*Coriandrum sativum*), red amaranthus (*A. paniculatus*) and green amaranthus (*A. frumentaceus*) using paper chromatography and their health implications. *J Herb Med Toxicol*, 4(1): 173–7.

Neffati, M., and Marzouk, B. 2008. Changes in essential oil and fatty acid composition in coriander (*Coriandrum sativum* L.) leaves under saline conditions. *Ind Crop Prod*, 28: 137–42.

Oganesyan, E. T., Nersesyan, Z. M., and Parkhomenko, A. Y. 2007. Chemical composition of the above ground part of *Coriandrum sativum*. *Pharm Chem J.*, 41(3): 30–4.

Padalia, R. C., Karki, N., Sah, A. N., and Verma, R. S. 2011. Volatile constituents of leaf and seed essential oil of *Coriandrum sativum* L. *J Essent Oil Bear Plants*, 14(5): 610–6.

Pandeya, K. B., Tripathi, I. P., Mishra, M. K., Dwivedi, N., Pardhi, Y., and Kamal, A. 2013. A critical review on traditional herbal drugs: an emerging alternative drug for diabetes. *Int J Org Chem*, 3: 1–22.

Pathan, A. R., Kothawade, K. A., and Logade, M. N. 2011. Anxiolytic and analgesic effect of *Coriandrum sativum* Linn. *Int J Res Pharm Chem*, 1(4): 1087–99.

Platel, K., and Srinivasan, K. 2004. Digestive stimulant actions of spices: a myth or reality? *Indian J Med Res*, 119: 167–79.

Properzi, A., Angelini, P., Bertuzzi, G., and Venanzoni, R. 2012. Some biological activities of essential oils. *Med Aromatic Plants*, 2(5):1–4.

Quintans-Jnior, L. J., Barreto, R. S. S., Menezes, P. P., Almeida, J. R. G. S., Viana, A. F. S. C., Oliveira, R. C. M. 2013. β-Cyclodextrin-complexed

(−)-linalool produces antinociceptive effect superior to that of (−)-linalool in experimental pain protocols. *Basic Clin Pharmacol Toxicol.*, 113(3): 167-72.

Rahimi, A. R., Babaei, S., Kambiz, M., Asad, R., and Sheno, A. 2013. Anthocyanin content of coriander leaves as affected by salicylic acid and nutrients application. *Int J Biosci*, 3(2): 141–5.

Ravizza, R., Gariboldi, M. B., Molteni, R., and Monti, E. 2008. Linalool, a plant-derived monoterpene alcohol, reverses doxorubicin resistance in human breast adenocarcinoma cells. *Oncol Rep*, 20(3): 625–30.

Reuter, J., Huyke, C., Casetti, F., Theek, C., Frank, U., Augustin, M. 2008. Anti-inflammatory potential of a lipolotion containing coriander oil in the ultraviolet erythema test. *J Dtsch Dermatol Ges*, 6: 847–51.

Ringman, J. M., Frautschy, S. A., Cole, G. M., Masterman, D. L., and Cummings, J. L. 2005. Potential role of the curry spice curcumin in Alzheimer's disease. *Curr Alzheimer Res*. 2: 131–136.

Rodriguez-Amaya, D. B. 1999b. A Guide to Carotenoids Analysis in Food. ILSI Press, Washington, DC.

Sahib, N. G., Anwar, F., Gilani, A. H., Hamid, A. A., Saari, N., and Alkharfy, K. M. 2013. Coriander (*Coriandrum sativum* L.): A potential source of high-value components for functional foods and nutraceuticals a review. *Phytother Res*, 27: 1439-1456.

Sriti, J., Wannes, W. A., Talou, T., Mhamdi, B., Cerny, M., and Marzouk, B. 2010. Lipid profile of Tunisian coriander (*Coriandrum sativum*) seed. *J Am Chem Soc*, 87: 395–400.

Tachibana, Y., Kikuzaki, H., Lajis, N. H., and Nakatani, N. 2001. Antioxidative activity of carbazoles from *Murraya koenigii* leaves. *J Agr Food Chem*, 49: 5589-94.

Tahraoui, A., El Hilaly, J., Israili, Z. H., and Lyoussi, B. 2007. Ethnopharmacological survey of plants used in traditional treatment of hypertension and diabetes in southeastern Morocco (Errachidia province). *J Ethnopharmacol*, 110(1): 105–17.

Thabrew, M. I., Dharmasiri, M. G., and Senaratne, L. 2003. Anti-inflammatory and analgesic activity in the polyherbal formulation Maharasnadhi Quathar. *J Ethnopharmacol*, 85(2–3): 261–7.

Ugulu, I., Baslar, S., Yorek, N., and Dogan, Y. 2009. The investigation and quantitative ethnobotanical evaluation of medicinal plants used around Izmir province, *Turkey. J Med Plant Res*, 3(5): 345–67.

Verma, A., Pandeya, S. N., Yadav, S. K., Singh, S., and Soni, P. 2011. A review on *Coriandrum sativum* (Linn.): an ayurvedic medicinal herb of happiness. *J Adv Pharm Healthc Res*, 1(3): 28–48.

Waheed, A., Miana, G. A., Ahmad, S. I., and Khan, M. A. 2006. Clinical investigation of hypoglycemic effect of *Coriandrum sativum* in type-2 (NIDDM) diabetic patients. *Pak J Pharmacol*, 23(1): 7–11.

Wangensteen, H., Samuelsen, A. B., Malterud, K. E. 2004. Antioxidant activity in extracts from coriander. *Food Chem,* 88: 293–7.

Zare-Shehneh, M., Askarfarashah, M., Ebrahimi, L., MoradiKor, N., Zare-Zardini, H., Soltaninejad, H. 2014. Biological activities of a new antimicrobial peptide from *Coriandrum sativum. Int J Biosci*, 4(6): 89–99.

Zhang, Z., Chang, Q., Zhu, M., Huang, Y., Ho, W. K. K., and Chen, Z. Y. 2001. Characterization of antioxidants presents in hawthorn fruits. *J Nutr Biochem*, 12: 144–152.

In: Coriander
Editor: Deepak Kumar Semwal

ISBN: 978-1-53616-483-1
© 2019 Nova Science Publishers, Inc.

Chapter 7

ESSENTIAL OIL OF CORIANDER: A SOURCE OF ANTIMICROBIAL AGENT

*Sonali Aswal, Ankit Kumar, Ashutosh Chauhan and Deepak Kumar Semwal**
Uttarakhand Ayurved University, Dehradun, India

ABSTRACT

Coriandrum sativum L. also called as Cilantro, Arab parsley, Chinese parsley and Kasbour is an annual herb belonging to the family Umbelliferae/ Apiaceae. It is mainly cultivated for its fruits and/or leaves, which are used for different purposes such as food, drugs and spice. The aim of this chapter was to review the composition and health benefits of coriander oil together with other applications. The antimicrobial activity of the essential oil is extensively included in this chapter. The literature review was carried out by searching on the electronic databases including PubMed, Scopus, and Google Scholar for studies focusing on the biological and pharmacological activities of different parts of coriander. The compositional analysis of coriander seeds essential oil showed the presence of linalool, camphor, geraniol, α–pinene, γ–terpinene, geranyl acetate and limonene as major constituents in which linalool is the most

---

*Corresponding Author's Email: dr_dks.1983@yahoo.co.in.

abundant compound. The essential oil was found to have strong antimicrobial activities against a variety of human pathogens. The differential effects of the essential oil suggest that it exert activities via different mechanisms or pathways, warranting further investigations. These findings contribute to the knowledge about essential oils as sources of potential antimicrobial agent and encourage further investigations.

Keywords: *Coriandrum sativum*, essential oil, antibacterial activity, Terpenes, Phenolics

INTRODUCTION

One of the oldest herbs that have been used for over 3,000 years for both culinary and medicinal purposes is *Coriandrum sativum* L. (family Umbelliferae or Apiaceae). This herb is most popularly known as coriander in most parts of the globe. This edible plant is non-toxic to humans. Its essential oil is consumed on a daily basis. The essential oils from coriander seeds and leaves have been extensively investigated for their chemical composition and biological activities including antimicrobial, antioxidant, hypoglycemic, hypolipidemic, anxiolytic, analgesic, anti-inflammatory, anti-convulsant and anti-cancer activities, among others. Owing to the diverse applications, the essential oil is thus used in different ways, viz., in foods and in pharmaceutical products as well as in the perfumes and lotions.

Coriander seed oil is widely used in the food, health, cosmetics, soft drinks and chocolate industries all-around of the world. It plays a major role in preserving the shelf-life of foods by preventing their spoilage. The fruits are considered carminative, diuretic, tonic, stomachic, refrigerant and aphrodisiac. They are also used as a condiment in the preparation of sausages, seasonings and cookies, and as a flavouring for alcoholic beverages. In the Indian traditional medicine, coriander was used to treat the disorders of digestive, respiratory and urinary systems, as it has diaphoretic, diuretic and stimulant activity. The dietary uses of the coriander essential oil are therefore helpful in maintaining good health.

Coriander is found to be rich in the volatile oil. It is widely used as a spice and folk remedy. Coriander is an important ingredient of curry powder widely used in world cuisines. Data reveals that the essential oil of coriander has anti-diabetic, laxative, carminative, diuretic, tonic, hypolipidemic, anticancer effects (Katar et al., 2016).

PROPERTIES OF CORIANDER FRUITS AND SEEDS

The literature reveals that the fruits of coriander have many beneficial properties. The fruit of coriander, exhibits various remedial properties, can be used for cooking and for children's digestive upset and diarrhoea. Moreover, it also has antidiabetogenic, antibacterial, antimicrobial, anti-inflammatory, hypolipidaemic, anticancerous, antimutagenic, and antioxidant effects as evidenced by various research studies (Sriti Eljazi et al., 2017).

Coriander fruits are good antioxidant and useful in the treatment of many disorders caused by oxidative stress such as neurodegenerative and cardiovascular diseases (Aćimović et al., 2016). The dried fruits are used as food ingredients, cosmetics, perfumery and drugs. The essential oil and various extracts from coriander fruits act as a sedative to relief from nervousness (Hani et al., 2015).

Coriander fruits are used in the preparation of sausages, alcoholic beverages, cosmetics and perfumery. Rich in essential oil, the dry fruits of coriander are used in the preparation of liqueur, sweets and the home remedy includes respiratory and digestive affections. The ripe fruits of coriander are widely and popularly used in infusion preparation as an analgesic, antispasmodic, febrifuge, a carminative and diuretic agent. Home remedy includes respiratory and digestive affections (Khalid, K. A. 2015). In Iran, coriander fruits are used in the pickle, curry powders, sausages, cakes, pastries, biscuits and buns (Sourmaghi et al., 2015).

Seeds of coriander are used in the perfumery, food, tobacco, soft and alcoholic beverages, and pharmaceutical industries. It is one of the important seed spices used globally to add taste, flavour and pungency in various food items. It is also one of the frequently used ingredients in the preparation of Ayurvedic medicines and is a traditional home therapy for different ailments viz., rheumatism, joint pain, gastrointestinal complaints, flatulence, indigestion, insomnia, convulsions, anxiety, and loss of appetite (Rashed and Darwesh, 2015). The immature fresh green herb and mature seed as of the spice coriander are used for flavouring purposes. Rich green colour and characteristic aroma and flavour of coriander are important for its market and culinary values (Sekhon et al., 2015).

Coriandrum sativum essential oil and the seed extract possess antibacterial, antioxidant, antidiabetic, anticancerous and antimutagenic activities (Hani et al., 2015). Coriander seeds are found to be rich in protein, starch and oil and can be used as a valuable diet for livestock (Davazdahemami, 2015). The seeds are also used as antispasmodic, appetizer, carminative, diuretic and to treat nausea, seasonal fever, convulsion, insomnia and anxiety. It is also used to treat cough, bronchitis, dysentery, diarrhoea, gout, rheumatism, intermittent fevers and as antiedemic and antiseptic. Traditionally, both leaves and fruits are used to relieve nervousness and insomnia. It is commonly used in the pickle, curry powders, sausages, cakes, pastries, biscuits and buns (Sourmaghi et al., 2015).

Ripe fruits of coriander are widely and popularly used in infusion preparation as an analgesic, antispasmodic, febrifuge, a carminative and diuretic agent. Home remedy includes respiratory and digestive affections (Khalid, 2015).

PROPERTIES OF CORIANDER LEAVES

The leaves of coriander have been used as an aphrodisiac while traditionally used against nervousness and insomnia. Its use in various preparations like pickle, curry powders, sausages, cakes, pastries, biscuits and buns is quite common (Sourmaghi et al., 2015). The essential oil obtained from the leaves is widely used in the food, health, cosmetics, soft drinks and chocolate industries all-around of the world (Shams et al., 2016).

Coriander essential oil is used in the perfumery, cosmetic, as well as in the pharmaceutical industry for its antimicrobial activities. Being good antioxidant coriander fruits are useful in the treatment of many disorders caused by oxidative stress such as inflammations, diabetes, cancer, neurodegenerative and cardiovascular diseases. It can be used in organic agriculture as a companion plant for insect biological control or its essential oil as a potential biocide (Aćimović et al., 2016). The essential oil of coriander has anti-diabetic, laxative, carminative, diuretic, hypolipidemic, anticancer effects (Katar et al., 2016). The leaves and seeds of coriander plant are shown in Figure 1.

Figure 1. Coriander Leaves (A) and Seeds (B).

COMPOSITION OF ESSENTIAL OIL

There are numerous reports available on the composition of essential oil of coriander obtained from different parts of the world. Different reports showed the composition differently which may be due to variation in the climate, season, part and harvesting time (Bogavac et al., 2015; Maroufpoor et al., 2016; Zamindar et al., 2016; Sriti Eljazi et al., 2017).

Composition of Leaves Oil

Sousa et al., (2016) reported linalool (39.78%), linalool oxide (27.33%), p-cymene (17.62), camphor (7.45), α-pinene (4.95%), camphene (1.49%) and β-pinene (1.38%) as major chemical compounds from coriander leaves essential oil. According to Arjun et al., (2017), the composition of essential oil obtained from the fresh leaves is given in Table 1.

Table 1. Chemical constituents of essential oil of coriander fresh leaves

Constituents	Area %	Constituents	Area %
1-Decene	0.27	9-Tetradecenal	7.97
Nonanal	0.52	Z,E-2,13-Octadecadien-1-ol	0.13
1-Nonanol	0.40	Octadecanal	0.26
Hexadecane	0.15	α-Campholenal	0.33
Decanal	8.76	cis-10-Heptadecenoic acid, methyl ester	1.39
2-Cyclohexen-1-ol	3.83	Hexahydrofarnesylacetone	0.52
Cyclooctane	8.32	1-Hexadecene	0.32
Pentylcyclopentane	0.20	Tetradecanal	0.18
10-Undecenal	0.25	Phytol	0.26
2-Octenal	4.54	Carbonic acid, tridecyl 2,2,2-trichloroethyl ester	0.43
2-Decenal	1.92	Tetratriacontyl pentafluoropropionate	0.38
Methylcyclohexane	0.76	Octatriacontyl trifluoroacetate	1.10
Dodecanal	8.56	Pentadecane	0.24
trans 2-Dodecenal	17.81	1-Heneicosanol	0.40
2,7,10,14- Tetramethyl tridecane	0.46	Tetradecane	1.96
Hexadecanal	0.63	4-Tert.butyl-1-methyl-1-cyclohexanol	0.46
4,5-Epoxy-2-decenal	0.89	Behenic alcohol	0.79
2-Methylenecyclopentanol	9.11	1-Heptadecene	0.63
Caryophyllene oxide	1.20	n-Tetracosanol-1	0.36
Dodecanoic acid	1.83	2-Dodecen-1-yl(-)succinic anhydride	0.57
Tridecanal	1.06	3-Heptadecene	0.67
2,2,4,4,6,8,8-Heptamethylnonane	0.48	1-Heneicosyl formate	1.22
2-Tridecenoic acid	4.11	Cyclooctacosane	0.18

Composition of Seeds/Fruits Oil

The major constituents reported in the Russian coriander essential oil are 1,2-dimethyl-3-(1-methylethenyl) cyclopentanol (0.14%), 2,6-dimethyl-3,7-octadiene-2,6-diol (0.57%), camphene (1.09%), camphor (7.94%), davanone (0.15%), farnesol (0.24%), geranial (0.11%), geranyl acetate (3.19%), geranyl formate (0.13%), limonene (3.29%), linalool (59.92%), p-cymene (7.44%), terpinene-4-ol (0.11%), trans-linalool oxide (7.70%), α-pinene (6.44%), α-terpineol (0.81%), β-myrcene (0.20%), β-pinene (0.34%) and γ-vinyl-γ-valerolactone (0.16%) (Choi and Lee, 2018). According to Hani et al., (2015), the seed oil contains anethole (0.46-8.20%), borneol (0.78-3.12%), camphor (0.73-3.54%), carvacrol (0.16-2.18%), decanal (0.0-1.43%), geranial (1.37-4.56%), geraniol (1.13-2.06%), limonene (0.59-2.32%), linalool (35.13-65.58%), nerol (3.52-22.36%), neryl acetate (0.28-1.07%), p-cymene (0.21-0.78%), thymol (0.35-1.46%), undecanal (0.26-0.98%), α-terpinene (0.65-3.74%), α-pinene (0.0-1.23%), α-terpineol (7.17-16.68%) and γ-terpinene (0.56-1.86%). According to Shrirame et al., (2018), the composition of seeds oil is given in Table 2.

Table 2. Chemical constituents of coriander seeds oil

Constituents	Area %	Constituents	Area %
(-)-Carvone	2.97	n-heneicosane	0.08
1-Phenyltridecane	0.31	n-Hexatriacontane	1.12
2,3-Cycloheptenopyridine	0.07	n-Nonacosane	0.69
2-Bromotetradecane	0.03	n-Pentadecanoic acid	0.44
2-Caren-10-al	0.28	n-Pentadecylic acid	15.3
2-Methylhexacosane	1.56	n-Tetratetracontane	0.6
2-Nonadecanone	0.07	n-Tridecanal	0.04
2-p-Cymenol	0.12	Ocemene	0.03
4-bromo-2-adamantanol	4.49	o-Isopropylphenetole	0.04
Amisol LDE	0.35	p-Cumenol	0.03
Arachic acid	0.2	p-Cymene	0.26
Capric acid	0.1	Petroselinic acid	47.58
cis-Dihydrocarvone	0.02	Stigmasterol	0.54

Table 2. (Continued)

Constituents	Area %	Constituents	Area %
cis-Isoapiole	3.89	Terpinen-4-ol	0.03
Crodacid	1.91	Trans-2-dodecenoic acid	0.23
Dioctadecyl thiodiacetate	0.3	Trans-2-undecenoic acid	0.84
D-Limonene	0.07	Trans-dihydrocarvone	0.17
Durenol	0.04	Trans-gernylgeraniol	1.3
Eicosane	0.1	Trans-Squalene	0.38
Ethyl linolate	0.2	Vanicol	0.5
Hystrene	0.07	Vanillin	0.03
Industrene R	0.43	α-Amyrine	0.38
Linalool	5.17	α-bisabolene	0.05
Methyl petroselinate	0.46	β-Alanine, trimethylsilyl ester	0.03
Monoolein	0.96	β-caryophyllene	0.03
Myristic acid	0.26	β-Citronellol	0.1
Myristicin	0.15	β-geraniol	0.36
Myristyl palmitate	0.03	β-Sesquiphellandrene	0.03
n-Dodecanal	0.02	γ-sitosterol	0.27
Nerylacetate	1.73	γ-Terpinene	0.21

PRE-TREATMENT, DIFFERENT EXTRACTION METHODS AND THEIR EFFECTS ON COMPOSITION OF ESSENTIAL OIL

Drying of Samples

A study was conducted to find out the effect of drying methods on the essential oil of the aerial parts of coriander. An Iranian landrace of coriander was dried in sunlight, shade, mechanical ovens (40 and 60°C), microwave oven (500 and 700 W), and by freeze–drying. For analyzing the essential oils, GC-FID and GC/MS were used. According to the experiment, freeze-dried tissue gave the highest yield of essential oil (0.18 mL/100 g DW) followed by the shade-dried sample (0.13 ml/100 g DW). The main components determined in the oils from fresh and dried coriander were decanal (0-37.5%), cis-phytol (1.0-34.1%), 1-tetradecanol (trace-31.7%), 2E-dodecenal (8.3-17.2%), dodecanal (0.5-14.8%), n-decanol (0.5-14.8%), trans-2-undecen-1-ol (trace-12.9%), 2E-decenal (0–11.3%), 1-eicosanol (0–6.4%), and methyl chavicol (0–6.0%) (Pirbalouti et al., 2017).

Enzyme Pre-Treatment

Abbassi et al., (2018) conducted a study to evaluate the effects of enzyme pre-treatment on yields and chemical composition of the essential oil of coriander seeds. Enzyme treatment of coriander seeds resulted in improved essential oil yield and its main component linalool together with camphor and geranyl acetate.

Microwave Extraction

A study by Huzar et al., (2018) was performed to investigate the effect of different factors on the yield and composition of coriander essential oil obtained by hydrodistillation. The study found linalool as a major component among 12 compounds identified by GC-MS. The study suggested that the yield and production of coriander essential oil can be enhanced in a shorter period of time by using microwave heating during hydro-distillation. According to this study, linalool (78.45%), camphor (3.90%), geranyl acetate (2.13%) and geraniol (1.07%) were the major constituents whereas α-pinene, camphene, β-pinene, β-myrcene, D-limonene, γ-terpinene and terpinolene were the minor ones.

Supercritical CO_2 Extraction

In a separate study, Shrirame et al., (2018) found that the supercritical CO_2 extraction method can improve the total yield of the oil up to 4.55% under optimized conditions. The study reported the major constituents in this essential oil were petroselinic acid (47.58%), n-pentadecylic acid (15.3%), linalool (5.17%), 4-bromo-2-adamantanol (4.49%) and cisisoapiole (3.89%). A study was conducted for the supercritical CO_2 extraction and characterization of Coriander essential oil.

In Romania, coriander seeds were harvested and minced in order to obtain a ground powder with a various particle size of 500, 630 and 710 μm. The best extraction yield was 0.57% w/w which was achieved for the 630-μm particle size ground at 40°C; for this temperature, the density of CO_2 is higher as well as its solvation power. Through gas chromatography, a maximum yield of coriander essential oil extract was obtained. It has been found that 10 compounds made up nearly 97% of the extract. The supercritical CO_2 extraction of coriander essential oils is an environmentally friendly technique that results in pure solvent-free extracts with high biological potential (Dima et al., 2016).

Conventional Hydro-Distillation Method

The essential oil of the aerial parts was extracted through the conventional hydro-distillation method and subjected to GC-MS to identify their components. Trans 2-dodecenal (17%), 2-methylenecyclopentanol (9%), dodecanal (8%), cyclooctane (8%), 9-tetradecenal (8%) and decanal (8%) were found to be the major components in coriander oil (Arjun et al., 2017). Misharina (2016) found linalool and γ-terpinene as major components of seed oil. Priyadarshi et al., (2016) reported n-decanol as major constituent followed by dodecen-1-ol, decanal, 13-tetradecenal, dodecanal, tridecen-1-al and tetradecanal in the foliage of two varieties (Vulgare alef and Microcarpum DC) of *Coriandrum sativum*.

Soxhlet Extraction

Coriander essential oil was isolated by hydrodistillation and solid-liquid extraction with methylene chloride. Comparison of the essential oil was done with green and environmentally safe processes such as supercritical fluid extraction and subcritical water extraction. Supercritical fluid extraction was performed with pressures (100 and 300 bar) and in case of supercritical water extraction, different temperatures were used (100, 150 and 200°C) while other extraction conditions were the same. The comparison was done in terms of total extraction yield and qualitative and quantitative composition. Using Soxhlet extraction, highest extraction yield (14.45%) was achieved followed by supercritical fluid extraction (8.88%) and supercritical water extraction (0.36%). The constituents identified in this study were γ-terpinene, (+)-limonene, linalool, camphor and geraniol (Pavlić et al., 2015).

Using a conventional extraction method, Iranian and European coriander showed 71% linalool composition in the oil, although European coriander had more seed essence percentage (300%) and seed essence yield (58%) than Iranian coriander (Davazdahemami, 2015).

EFFECT OF FERTILIZERS AND OTHER FACTORS ON THE YIELD OF CORIANDER OIL

A study was conducted to find out the effect of weather conditions, location and fertilization on coriander fruit essential oil quality. Coriander was grown on three different localities and applied with six different fertilization regimes. Linalool (73.23%) was found to be the main component followed by α-pinene (8.43%), γ terpinene (7.64%), camphor (3.07%), limonene (1.78%) and geranyl acetate (1.57%), while other compounds (camphene, p-cymene, β-pinene, myrcene and sabinene) were present in less than 1%. The weather conditions strongly influenced the compound content while the application of different fertilizers significantly influenced the content of linalool, α-pinene, γ-terpinene and limonene. The study suggested that growing coriander in a dry climate is more favourable for linalool accumulation in essential oil than moderate to humid climate.

Moreover, fertilization positively influences linalool content in coriander essential oil (Milica et al., 2016).

Coriander was studied to find out the effect of microclimate on its planting date and water requirements under different nitrogen sources on growth, yield, oil yield and components of the plants. Nitrogen fertilization on growth, yield and components of *Coriandrum sativum* plants were used. The components identified in coriander oil were α-pinene, myrcene, camphene, β-pinene, p-cymene, limonene, borneol, linalool, nerol, menthone, geraniol, eugenol, geranyl acetate and terpinene-4-ol. Under all planting dates with all fertilization treatments, the highest percentage was of linalool (77.31%) (Rashed and Darwesh, 2015).

Hani et al., (2015) conducted a study to investigate the effect of phosphorus fertilization (0, 12 and 24 kg-ha^{-1} P) on the seed yield, and volatile oil and chemical composition of coriander under water stress conditions. The study suggested that coriander plants have high water use efficiency at 50% evapotranspiration crop, and 24 kg-ha^{-1} P whilst irrigation at 100% evapotranspiration crop without any phosphorus application gave the lowest water use efficiency. Concerning essential oil constituents, linalool was the major compound in coriander fruits.

Aćimović et al., (2015) evaluated the influence of the application of various types of fertilizers on essential oil content in coriander fruits by applying different types of fertilizer approved for the organic farming system, as well as mineral fertilizer. The application of vermicompost, Royal Ofert biohumus and chemical fertilizer gave the highest content of coriander essential oil. A separate study by Khalid (2015) investigated the effect of different levels of nitrogen and phosphorous fertilizers, trace elements and their interactions on the essential oil extracted from Egyptian coriander plants. The main constituents identified from the oil were linalool (75.5%-75.8%), limonene (6.8%-7.3%) and camphor (3.7%-4.3%) which increased as Nitrogen and Phosphorus levels increase. Application of nitrogen and phosphorous micronutrients caused a pronounced increment in both essential oil content and yield of coriander compared with the treatments of nitrogen and phosphorous without micronutrients.

Effect of foliar application of K as potassium nitrate ability to mitigate the negative impacts of salinity on coriander plants was also investigated by Elhindi et al., (2016). In a greenhouse condition 0, 40 and 80 mM of NaCl were applied to the growth medium while 0, 50 and 100 mM of K as KNO_3 was adjusted two times as a foliar spray on the plants. The essential oil production of coriander plants was significantly reduced under salinity and slightly increased under KNO_3. The coriander oil production was decreased significantly under salinity stress conditions.

STUDY ON DIFFERENT VARIETY OF CORIANDER

A study was conducted by Beyzi et al., (2017) which focused on to the study of crude oil content, fatty acid-mineral composition and bioactive properties of seed oil extracted from different coriander varieties namely Arslan, Gürbüz, Gamze and Erbaa. Essential oil composition was also monitored in these varieties and compared. The highest crude oil content was calculated for the Gamze cultivar while the lowest was in Arslan variety. Linalool was identified as the main component, followed by camphor and geraniol. Linalool ratios ranged from 89.44% to 91.77% and it was the highest in Erbaa variety and the lowest was in Gurbuz variety. The lowest antiradical activity was found in the Erbaa cultivar and the highest amount of total phenolic compounds was for the Arslan cultivar.

A research was conducted to examine the effects of two different ecologies in the essential oil composition of new Turkish coriander cultivar. Turkish coriander cultivars, Gamze, Arslan, Erbaa, Pelmus, Kudret and Gürbüz were examined in Mardin and Tokat. Essential oil contents were high in Erbaa, Pelmus and Kudret-cultivars. Linalool was the component in all the cultivars. Maximum linalool was obtained in Arslan and Erbaa cultivars with 92.1% and 90.4%, respectively. Contrary to linalool, α-pinene and neryl acetate contents of all cultivar were higher in Tokat (İzgı et al., 2017).

A quantitative study was conducted to compare coriander essential oil and dry matter yield in different climates of Iran. Cultivation of coriander seeds was carried out in pots with four replications in the cities of Maku, Khoy and Urmia with an altitude of 1182 m, 1148 m and 1332 m, respectively. Environmental factors significantly affected the oil content and the dry matter yield of coriander. The fruits of Maku and Khoy gave the maximum oil content while Urmia gave the minimum oil content. The oil content and dry matter yield increased with decreasing altitude and precipitation in the cities of Maku and Khoy. The essential oil content of coriander fruits, collected from Makou and Khoy, was identical because of the almost identical climate and environmental conditions of these two cities. It can be concluded that among the climate parameters, altitude and precipitation have the most impact on the essential oil content and dry matter yield of coriander. Low altitudes could increase the essential oil content of coriander (Shams et al., 2016). Katar et al., (2016) performed a study using Arslan, Erbaa, Gamze, Kudret-K and Pel-Mus which are the five different cultivars along with three different genotypes Burdur, Antalya and Tokat. In essential oil content, Pel-mus and Erbaa genotypes had the highest essential oil content. Moreover, Burdur genotype had the lowest essential oil content. Different climatic and geographic characters had significant effects in coriander genotypes for yield, yield-components and essential oil content.

A two-year study was conducted by Unlukara et al., (2016) to determine the essential oil contents and composition of *Coriandrum sativum* cultivars across different irrigation levels in a semiarid climate area of Kayseri, Turkey. *C. sativum* varieties Aslan and Gurbuz were evaluated under 5 different water stress conditions. To create water stress, 100%, 75%, 50% and 25% of depleted water from field capacity were applied. Coriander is a slightly sensitive plant to water stress since seed yield response factor of both coriander cultivars were found slightly higher than 1. Higher seed and essential oil yields were obtained with elevated water application. Water stress caused a decrease in coriander essential oil yield. Essential oil ratios were not changed significantly by water stress. Main essential oil component linalool was not changed with changing water amounts. Non-stable differences in essential oil components were observed between coriander varieties in both years. In coriander cultivation, timely irrigation with enough water is very crucial to obtain the highest seed yield and essential oil yield.

ANTIMICROBIAL ACTIVITY OF CORIANDER ESSENTIAL OIL

The leaf oil was evaluated for its antimicrobial activity against the oral pathogens *Candida albicans, Fusobacterium nucleatum, Porphyromonas gingivalis, Streptococcus sanguis* and *Streptococcus mitis*. The MIC values for *F. nucleatum* and *S. mitis* were recorded to 0.015 and 0.062 mg/mL, respectively. All oral species were inhibited by the essential oil with MICs ranging from 0.007 to 0.250 mg/mL, and minimal bactericidal/ fungicidal concentrations from 0.015 to 0.500 mg/mL. The GCMS studies revealed the presence of alcohol and aldehyde derivatives in the oil (Bersan et al., 2014). The essential oils and bioactive fractions from *Coriandrum sativum* prove to have strong antimicrobial activity on planktonic microorganisms. Confocal analysis was used to investigate the effect of the essential oil and bioactive fraction on the morphology of *S. mutans* biofilms (thickness, biovolume, and architecture) and on the metabolic viability of *C. albicans* biofilms. *C. sativum* oil drastically affected *C. albicans* viability when compared to the control. The essential oil also inhibits yeast biofilm adherence onto a polystyrene substrate from 62.5 μg/mL. Confocal microscopic analysis revealed that the essential oil notably altered the viability of yeast cells compared to the vehicle which considerably decreased the metabolic activity of the fungal cells. Decanal and trans-2-decenal are the major compound identified in *C. sativum* essential oil having strong fungicidal effect against *Candida albicans* and *non-albicans* and acts by binding to membrane ergosterol, which increases ionic permeability and leads to cell death. The study suggested that coriander oil emerges as a promising candidate for nonclinical and clinical toxicology testing for the development of new drugs to treat denture-related oral candidiasis (Freires et al., 2015). The oil having linalool as a major constituent was found to have antimicrobial activity against *B. subtilis, C. albicans, E. faecalis, E. faecium, K. pneumonia, L. innocua, P. aeruginosa, S. enteritidis, S. infantis, S. kentucky*, and *S. typhimurium* with the MIC values ranging between <0.195 and 1.562 µg/mL. Moreover, the oil showed a moderate antimicrobial activity against *E. aerogenes, L. monocytogenes, P. fluorescens, S aureus*, and *S. epidermidis* with the MIC values between 25 and 3.125 µg/mL (Özkinali et al., 2017).

The oil from the fruits was tested for its antimicrobial activity against *Escherichia coli, Salmonella* sp. (clinical isolate), *Staphylococcus aureus, Proteus vulgaris, Bacillus cereus, Penicillium spp., Rhizopus spp., Aspergillus niger* and *Saccharomyces cerevisiae* by disc-diffusion and well-diffusion methods. The extracts from fruits of coriander inhibited the microorganisms with IZD values between 8 and 18 mm while the minimum inhibitory concentration of more than 600 ppm. On the other hand, the oils from fruits inhibited with IZD values of 8-10 mm and MIC of more than 600 ppm. The study suggested that the oil, as well as the extract, can be useful as a bio-preservative agent (Teneva et al., 2015).

Against the clinical vaginal bacterial Gram-negative strains (*Escherichia coli, Proteus mirabilis*), Gram-positive strains (*Staphylococcus aureus, Enterococcus sp.*) and clinical yeast strain (*Candida albicans*), the essential oil showed activity with MIC ranged from 0.4 to 45.4 µL/mL, while the MBC values were recorded between 0.8 and 45.4 µL/mL. *E. coli* showed the highest susceptibility to the essential oil (MIC = 0.4 µL/mL) while much lower activity was obtained for the strain *S. aureus* (MIC = 2.8 µL/mL). The Brine shrimp lethality bioassay revealed that the oil is non-toxic as it showed LC_{50} value of 2.25 mg/mL (Bogavac et al., 2015). The coriander oil produced inhibition halos between 17 and 33.7 mm, and between 18.4 and 25 mm against *Staphylococcus epidermidis* and *Staphylococcus aureus*, respectively. The study found that linalool was the responsible constituent for antibacterial activity (Scazzocchio et al., 2016).

Antifungal Activity

The leaf oil showed antifungal activity against *Candida spp* with MIC values ranged from 15.6 to 31.2 mg/mL, and MFC values ranged from 31.2 to 62.5 mg/mL. The crude essential oil was shown to have a better antifungal effect than its active fraction. Antifungal properties of *C. sativum* essential oil are not related to cell wall biosynthesis pathways because of the presence or absence of an osmotic protector (sorbitol). The essential oil and active fraction from *C. sativum* L. had low cytotoxicity on human HeLa cells with IC$_{30}$ of 359.76 mg/mL (de Almeida Freires et al., 2014). The oil was also found activity against *Aspergillus parasiticus, Cladosporium cladosporioides, Eurotium herbariorum, Penicillium chrysogenum* and *Aspergillus carbonarius* with the effective concentration at 4.17 µL/mL (Dimić et al., 2015). Coriander seeds essential oil had a significant effect on mycelial growth of *Colletotrichum acutatum* at ≥0.16 µL/mL (Aćimović et al., 2016). Sousa et al., (2016) found that leaves oil showed activity against *Candida albicans* isolated from human faeces. The essential oil showed MIC and MFC values of 512 and 1024 µg/mL, respectively.

Antibacterial Activity

Alboofetileh et al., (2014) found that coriander oil partially inhibited *Listeria monocytogenes* while poorly inhibited *Staphylococcus aureus* and *Escherichia coli*. The activity of fruit oil against the food-borne pathogens comprising *Staphylococcus aureus, Listeria monocytogenes, Escherichia coli, Salmonella typhimurium* and *Bacillus cereus* has been recorded in which *Staphylococcus aureus* displayed the highest susceptibility towards coriander oil. This oil was found to contain linalool (71.8%) and camphor (4.3%) (El-Shenawy et al., 2015). The oil was found active against *Bacillus cereus, Staphylococcus aureus, Salmonella typhimurium* and *E. coli*. Among tested bacterial species, *Bacillus cereus* was found most sensitive for oil exhibiting the highest percentage of 97%. The inhibition percentage was 99% in the case of *Salmonella typhimurium* when coriander oil was used at the concentration of 10 μL/mL. At a similar concentration, it gave the highest effect on *E. coli* by 99% inhibition. Coriander oil showed complete inhibition of *Salmonella typhimurium, E. coli* and *S. aureus* when it was used at the concentration of 30 μL per well. The antibacterial effect of the coriander essential oil gave the highest antibacterial effect on all strains with inhibition percentages ranged from 90-99% and the inhibition zone ranged between 80-90 mm in diameters (Mohamed et al., 2016).

Bazargani and Rohloff (2016) investigated in vitro antibiofilm activities of essential oil and extracts of coriander seeds. The coriander oil displayed the highest inhibitory activity as compared to the other extracts against *S. aureus* and *E. coli*. The results indicated that essential oil of coriander could inhibit biofilm formation of *E. coli* completely, displaying 100% inhibition activities. Coriander oil had the highest inhibitory potential with 86% and 71.5% reduction in metabolic activity of *E. coli* and *S. aureus*, respectively. Dima et al., (2016) investigated the antibacterial activity of oil on three food-borne pathogens *Escherichia coli, Salmonella spp.* and *Staphylococcus aureus*. The coriander essential oil significantly inhibited the growth of pathogens when observed after 10 days at 10°C. The coriander essential oil obtained using the supercritical CO_2 extraction demonstrated an antibacterial activity, which qualifies its use both as a condiment and as a preservative.

A study was carried out to determine the effect of seeds oil on the growth of selected bacteria strains of the *Lactobacillus* genus. The zone size was recorded at ranged between 0.1 and 5.8 mm. The oil concentration of 50% inhibited the growth of all the tested strains of lactic acid bacteria while less than 50% concentration affected only some selected bacterial strains of the *Lactobacillus* genus (Kozłowska et al., 2018). The antibacterial activity of coriander essential oil and its major constituent, linalool was tested with antibiotics against methicillin-susceptible and methicillin-resistant *Staphylococcus aureus, S. epidermidis, Pseudomonas aeruginosa* and *Escherichia coli*. Coriander essential oil and linalool inhibited bacterial growth with MIC values of 5.44 and 5.36 μg/mL, respectively. Other constituents detected in coriander essential oil (α-pinene, camphor, γ-terpinene, geranyl acetate and D-limonene) have also antibacterial effects (Aelenei et al., 2019).

The seed and leaf of coriander were evaluated for their possible synergistic interactions on antibacterial and antioxidant efficacy of essential oils in combination. Antibacterial combination effect was evaluated against *Bacillus cereus, Listeria monocytogenes, Micrococcus luteus, Staphylococcus aureus, Escherichia coli* and *Salmonella typhimurium* using micro broth dilution, checkerboard titration and time-kill methods. Results showed that coriander in combination with cumin seed oil showed synergistic interactions both in antibacterial (FICI = 0.25-0.50) and antioxidant (CI = 0.79) activities. The bioactive compound from coriander seed oil was identified as linalool (68.69%). The coriander/cumin seed oil combination did not show any cytotoxic effect both in brine shrimp lethality as well as human normal colon cell line assays. Coriander/cumin seed oil combination might indeed be used as a potential source of safe and effective natural antimicrobial and antioxidant agents in pharmaceutical and food industries (Bag and Chattopadhyay, 2015). The oil also exhibited antibacterial activity against multidrug-resistant *Escherichia coli* with MIC value of 0.6 mg/mL (Mansouri et al., 2018). The oil was also found able to inhibit the growth of uropathogenic *Escherichia coli* and found to reduce the MIC of gentamicin (Scazzocchio et al., 2017). Sourmaghi et al., (2015) observed that the oil obtained by hydro-distillation showed stronger antimicrobial activity against *Staphylococcus aureus* and *Candida albicans* than that of the oil obtained by a microwave-assisted method with inhibition of 63% and 66%, respectively. Hydro-distilled oil showed greater activity MIC against *E. coli* and *P. aeruginosa* by 0.781 and 6.25 µL/mL, respectively. Gayathri et al., (2016) found that the coriander extract has an inhibitive effect against *Escherichia coli* and *Pseudomonas aeruginosa* but not against *Staphylococcus aureus*.

ANTIOXIDANT PROPERTIES

The antioxidant potential of essential oil of coriander in various *in vitro* models and in food supplements enriched with omega-6 and omega-3 fatty acids were evaluated and compared. Coriander oil showed significantly higher radical scavenging and Fe^{2+} ion-chelating potential. Strong radical scavenging and Fe^{2+}-chelating, as well as anti-lipid peroxidative activities of coriander oils, suggest that it serve as a potential source of natural antioxidants for retarding lipid oxidation of food supplements enriched with omega-6 and omega-3 fatty acids (Bag and Chattopadhyay, 2018). Seeds essential oil was studied for its antioxidant activity. The highest antioxidant activity was observed with nitric oxide scavenging (0.28 mg/100g) and reducing power activity (0.27 mg/100g). The study suggested that coriander seeds essential oil could be used as a raw material for the preparation of pharmacologically active products (Krishnaveni and Santhoshkumar, 2016).

ACTIVITY AGAINST ALZHEIMER'S DISEASE

Cioanca et al., (2014) conducted a study to analyze the anxiolytic and antidepressant activity coriander volatile oil. Effects of inhaled coriander volatile oil on anxiety and depression levels were studied. The possible anxiolytic, antidepressant and antioxidant propriety in a beta-amyloid rat model of Alzheimer's disease were analyzed. Wistar rats were treated with coriander oil 1% and 3%. Rats were exposed to oil vapours for controlled 60 min period, daily, for 21 continuous days. Coriander volatile oil decreased catalase activity and increased glutathione level in the hippocampus. Coriander volatile oil counteracts with anxiety, depression and oxidative stress in Alzheimer's disease conditions. There is no toxic or harmful effect of coriander volatile oil with respect to its dietary use hence it is considered safe to use.

INSECTICIDAL ACTIVITY

The essential oil of *Coriandrum sativum* seeds was used to compare its insecticidal activity against *Sitophilus oryzae* and *S. zeamais*. The essential oil extracted by steam distillation extraction showed the highest insecticidal activity against *S. oryzae* and *S. zeamais* (Choi and Lee, 2018). Recently, Amini et al., (2018) performed a study to evaluate the biological effects of the four different species of Apiaceae family against adult insects of two important storage pests *Sitophilus oryzae* and *Tribolium castaneum*. Coriander oil showed the most fumigant toxicity on the storage pests. A separate study by Sriti Eljazi et al., (2017) was conducted to evaluate the insecticidal activity of coriander essential oil and its major compound, linalool, against three stored product insects *Tribolium castaneum, Lasioderma serricorne* and *Sitophilus oryzae*. Chemical analysis indicated that linalool was the main component of coriander essential oil. Coriander essential oil was relatively more toxic against *T. castaneum, L. serricorne*, and *S. oryzae*. The increasing doses of essential oils caused a significant increase in the mortality when the *T. castaneum, L. serricorne*, and *S. oryzae* adults were exposed to the oil for 72 h. Percentage mortality of *T. castaneum, S. oryzae*, and *L. serricorne* were 66.67%, 70%, and 100%, respectively, at the 625 µL/L air dose after 24 h exposure when exposed to essential oil of coriander and the same dose caused complete mortality (100%) when exposed to linalool against *T. castaneum* and *L. serricorne* and 80% against *S. oryzae*. This study suggested that coriander oil has the potential for use in insecticidal activities.

REFERENCES

Abbassi, A., Mahmoudi, H., Zaouali, W., M'rabet, Y., Casabianca, H., & Hosni, K. (2018). Enzyme-aided release of bioactive compounds from coriander (Coriandrum sativum L.) seeds and their residue by-products and evaluation of their antioxidant activity. *Journal of Food Science and Technology*, *55*(8), 3065-3076.

Aćimović, M. G., Grahovac, M. S., Stanković, J. M., Cvetković, M. T., & Maširević, S. N. (2016). Essential oil composition of different coriander (Coriandrum sativum L.) accessions and their influence on mycelial growth of Colletotrichum spp. *Acta Scientiarum Polonorum - Hortorum Cultus*, *15*(4), 35-44.

Aelenei, P., Rimbu, C. M., Guguianu, E., Dimitriu, G., Aprotosoaie, A. C., Brebu, M., Horhogea, C. E. & Miron, A. (2019). Coriander essential oil and linalool–interactions with antibiotics against Gram-positive and Gram-negative bacteria. *Letters in Applied Microbiology*, *68*(2), 156-164.

Alboofetileh, M., Rezaei, M., Hosseini, H., & Abdollahi, M. (2014). Antimicrobial activity of alginate/clay nanocomposite films enriched with essential oils against three common foodborne pathogens. *Food Control*, *36*(1), 1-7.

Amini, S., Tajabadi, F., Khani, M., Labbafi, M. R., & Tavakoli, M. (2018). Identification of the Seed Essential Oil Composition of Four Apiaceae Species and Comparison of their Biological Effects on Sitophilus oryzae L. and Tribolium castaneum (Herbst.). *Journal of Medicinal Plants*, *3*(67), 68-76.

Arjun, P., Kumar Semwal, D., Badoni Semwal, R., Malaisamy, M., Sivaraj, C., & Vijayakumar, S. (2017). Total Phenolic Content, Volatile Constituents and Antioxidative Effect of Coriandrum sativum, Murraya koenigii and Mentha arvensis. *The Natural Products Journal*, *7*(1), 65-74.

Bag, A., & Chattopadhyay, R. R. (2015). Evaluation of synergistic antibacterial and antioxidant efficacy of essential oils of spices and herbs in combination. *PloS one*, *10*(7), e0131321.

Bag, A., & Chattopadhyay, R. R. (2018). Evaluation of antioxidant potential of essential oils of some commonly used Indian spices in in vitro models and in food supplements enriched with omega-6 and omega-3 fatty acids. *Environmental Science and Pollution Research*, *25*(1), 388-398.

Bazargani, M. M., & Rohloff, J. (2016). Antibiofilm activity of essential oils and plant extracts against Staphylococcus aureus and Escherichia coli biofilms. *Food Control*, *61*, 156-164.

Bersan, S. M., Galvão, L. C., Goes, V. F., Sartoratto, A., Figueira, G. M., Rehder, V. L., Alencar, S. M., Duarte, R. M., Rosalen, P. L., & Duarte, M. C. (2014). Action of essential oils from Brazilian native and exotic medicinal species on oral biofilms. *BMC Complementary and Alternative Medicine*, *14*(1), 451.

Beyzi, E., Karaman, K., Gunes, A., & Beyzi, S. B. (2017). Change in some biochemical and bioactive properties and essential oil composition of coriander seed (Coriandrum sativum L.) varieties from Turkey. *Industrial Crops and Products*, *109*, 74-78.

Bogavac, M., Karaman, M., Janjušević, L., Sudji, J., Radovanović, B., Novaković, Z., Simeunović, J., & Božin, B. (2015). Alternative treatment of vaginal infections–in vitro antimicrobial and toxic effects of Coriandrum sativum L. and T hymus vulgaris L. essential oils. *Journal of Applied Microbiology*, *119*(3), 697-710.

Choi, S.-A., & Lee, H.-S. (2018). Insecticidal activities of Russia coriander oils and these constituents against Sitophilus oryzae and Sitophilus zeamais. *Journal of Applied Biological Chemistry*, *61*(3), 239-243.

Cioanca, O., Hritcu, L., Mihasan, M., Trifan, A., & Hanciann, M. (2014). Inhalation of coriander volatile oil increased anxiolytic–antidepressant-like behaviors and decreased oxidative status in beta-amyloid (1–42) rat model of Alzheimer's disease. *Physiology & Behavior*, *131*, 68-74.

Davazdahemami, S. (2015). Comparison of Essential Oil Yield and Essential Oil Compositions of Iranian and European Corianders (Coriandrum sativum L.). *Journal of Essential Oil Bearing Plants*, *18*(3), 633-636.).

de Almeida Freires, I., Murata, R. M., Furletti, V. F., Sartoratto, A., de Alencar, S. M., Figueira, G. M., de Oliveira Rodrigues, J. A., Duarte, M. C. & Rosalen, P. L. (2014). Coriandrum sativum L. (coriander) essential oil: antifungal activity and mode of action on Candida spp., and molecular targets affected in human whole-genome expression. *PLoS One*, *9*(6), e99086.

Dima, C., Ifrim, G. A., Coman, G., Alexe, P., & Dima, Ş. (2016). Supercritical CO2 Extraction and Characterization of C oriandrum Sativum L. Essential Oil. *Journal of Food Process Engineering*, *39*(2), 204-211.

Dimić, G., Kocić-Tanackov, S., Mojović, L., & Pejin, J. (2015). Antifungal activity of lemon essential oil, coriander and cinnamon extracts on foodborne molds in direct contact and the vapor phase. *Journal of Food Processing and Preservation*, *39*(6), 1778-1787.

Duarte, A., Luís, Â., Oleastro, M., & Domingues, F. C. (2016). Antioxidant properties of coriander essential oil and linalool and their potential to control Campylobacter spp. *Food Control*, *61*, 115-122.

Elhindi, K. M., El-Hendawy, S., Abdel-Salam, E., Schmidhalter, U., ur Rahman, S., & Hassan, A. A. (2016). Foliar application of potassium nitrate affects the growth and photosynthesis in coriander (Coriander sativum L.) plants under salinity. *Progress in Nutrition*, *18*(1), 63-73.

El-Shenawy, M. A., Baghdadi, H. H., & El-Hosseiny, L. S. (2015). Antibacterial activity of plants essential oils against some epidemiologically relevant food-borne pathogens. *The Open Public Health Journal*, *8*(1), 30-34.

Farooq, M., Hegde, R. V., & Imamsaheb, S. J. (2017). Variability, heritability and genetic advance in coriander genotypes. *Plant Archives*, *17*(1), 519-522.

Freires, I. A., Bueno-Silva, B., Galvão, L. C. D. C., Duarte, M. C. T., Sartoratto, A., Figueira, G. M., de Alencar, S. M., & Rosalen, P. L. (2015). The effect of essential oils and bioactive fractions on Streptococcus mutans and Candida albicans biofilms: A confocal analysis. *Evidence-Based Complementary and Alternative Medicine*, *2015*.

Gayathri, P. K., Rithika, J., Dhanasree, S. (2016). Preliminary Phytochemical analysis and anti-microbial evaluation of the Cilantro extract. *Journal of Chemical and Pharmaceutical Sciences*, *9* (3), 1633-1637.

Han, X., Beaumont, C., & Stevens, N. (2017). Chemical composition analysis and in vitro biological activities of ten essential oils in human skin cells. *Biochimie open*, *5*, 1-7.

Hani, M. M., Hussein, S. A. A. H., Mursy, M. H., Ngezimana, W., & Mudau, F. N. (2015). Yield and essential oil response in coriander to water stress and phosphorus fertilizer application. *Journal of Essential Oil Bearing Plants*, *18*(1), 82-92.

Huzar, E., Dzięcioł, M., Wodnicka, A., Örün, H., İçöz, A., & Çiçek, E. (2018). Influence of hydrodistillation conditions on yield and composition of coriander (Coriandrum sativum L.) Essential oil. *Polish Journal of Food and Nutrition Sciences*, *68*(3), 243-250.

İzgı, M. N., Telci, İ., & Elmastaş, M. (2017). Variation in essential oil composition of coriander (Coriandrum sativum L.) varieties cultivated in two different ecologies. *Journal of Essential Oil Research*, *29*(6), 494-498.

Katar, D., Kara, N., & Katar, N. (2016). Yields and quality performances of coriander (Coriandrum sativum L.) genotypes under different ecological conditions. *Turkish Journal of Field Crops*, *21*(1), 78-86.

Khalid, K. A. (2015). Effect of macro and micro nutrients on essential oil of coriander fruits. *Journal of Materials and Environmental Science*, *6*(8), 2060-2065.

Kozłowska, M., Ziarno, M., Rudzińska, M., Tarnowska, K., Majewska, E., Kowalska, D. (2018). Chemical composition of coriander essential oil and its effect on growth of selected lactic acid bacteria. *Żywność. Nauka. Technologia. Jakość.* 25, 1 (114), 97-111.

Krishnaveni, M., & Santhoshkumar, J. (2016). Secondary Metabolite, Antioxidant, Phyto Nutrient Assay of Essential Oil from Dry Coriandrum sativum Seed Black Variety. *Research Journal of Pharmacy and Technology*, *9*(7), 853.

Mansouri, N., Aoun, L., Dalichaouche, N., & Hadri, D. (2018). Yields, chemical composition, and antimicrobial activity of two Algerian essential oils against 40 avian multidrug-resistant Escherichia coli strains. *Veterinary World*, *11*(11), 1539.

Maroufpoor, M., Ebadollahi, A., Vafaee, Y., & Badiee, E. (2016). Chemical composition and toxicity of the essential oil of Coriandrum sativum L. and Petroselinum crispum L. against three stored-product insect pests. *Journal of Essential Oil Bearing Plants, 19*(8), 1993-2002.

Milica, A., Mirjana, C., & Jovana, S. (2016). Effect of weather conditions, location and fertilization on coriander fruit essential oil quality. *Journal of Essential Oil Bearing Plants, 19*(5), 1208-1215.

Misharina, T. A. (2016). Antiradical properties of essential oils and extracts from coriander, cardamom, white, red, and black peppers. *Applied Biochemistry and Microbiology, 52*(1), 79-86.

Mohamed, H. G., Gaafar, A. M., & Soliman, A. S. (2016). Antimicrobial Activities of Essential Oil of Eight Plant Species from Different Families against some Pathogenic Microorganisms. *Research Journal of Microbiology, 11*(1), 28.

Özkinali, S., Şener, N., Gür, M., Güney, K., & Olgun, Ç. (2017). Antimicrobial activity and chemical composition of Coriander & Galangal essential oil. *Indian Journal of Pharmaceutical Education and Research, 51*(3), 221-224.

Pavlić, B., Vidović, S., Vladić, J., Radosavljević, R., & Zeković, Z. (2015). Isolation of coriander (Coriandrum sativum L.) essential oil by green extractions versus traditional techniques. *The Journal of Supercritical Fluids, 99*, 23-28.

Pirbalouti, A. G., Salehi, S., & Craker, L. (2017). Effect of drying methods on qualitative and quantitative properties of essential oil from the aerial parts of coriander. *Journal of Applied Research on Medicinal and Aromatic Plants, 4*, 35-40.

Priyadarshi, S., Khanum, H., Ravi, R., Borse, B. B., & Naidu, M. M. (2016). Flavour characterisation and free radical scavenging activity of coriander (Coriandrum sativum L.) foliage. *Journal of Food Science and Technology, 53*(3), 1670-1678.

Rashed, N. M., & Darwesh, R. K. (2015). A comparative study on the effect of microclimate on planting date and water requirements under different nitrogen sources on coriander (Coriandrum sativum, L.). *Annals of Agricultural Sciences, 60*(2), 227-243.

Scazzocchio, F., Garzoli, S., Conti, C., Leone, C., Renaioli, C., Pepi, F., & Angiolella, L. (2016). Properties and limits of some essential oils: chemical characterisation, antimicrobial activity, interaction with antibiotics and cytotoxicity. *Natural Product Research, 30*(17), 1909-1918.

Scazzocchio, F., Mondì, L., Ammendolia, M. G., Goldoni, P., Comanducci, A., Marazzato, M., Conte, M. P., Rinaldi, F., Crestoni, M. E., Fraschetti, C., & Longhi, C. (2017). Coriander (Coriandrum sativum) Essential Oil: Effect on Multidrug Resistant Uropathogenic Escherichia coli. *Natural Product Communications, 12*(4), 1934578X1701200438.

Shams, M., Ramezani, M., Esfahan, S. Z., Esfahan, E. Z., Dursun, A., & Yildirim, E. (2016). Effects of climatic factors on the quantity of essential oil and dry matter yield of coriander (Coriandrum sativum L.). *Indian Journal of Science and Technology, 9*(6), 1-4.

Shrirame, B. S., Geed, S. R., Raj, A., Prasad, S., Rai, M. K., Singh, A. K., Singh, R. S. & Rai, B. N. (2018). Optimization of Supercritical Extraction of Coriander (Coriandrum sativum L.) Seed and Characterization of Essential Ingredients. *Journal of Essential Oil Bearing Plants, 21*(2), 330-344.

Sourmaghi, M. H. S., Kiaee, G., Golfakhrabadi, F., Jamalifar, H., & Khanavi, M. (2015). Comparison of essential oil composition and antimicrobial activity of Coriandrum sativum L. extracted by hydrodistillation and microwave-assisted hydrodistillation. *Journal of Food Science and Technology, 52*(4), 2452-2457.

Sousa, J. P., Queiroz, E. O., Guerra, F. Q., Mendes, J. M., Pedrosa, Z. V., Pereira, F. O., Trajano, V. N., Souza, F. S., & Lima, E. O. (2016). Morphological alterations and time-kill studies of the essential oil from the leaves of Coriandrum sativum L. on Candida albicans. *Boletín Latinoamericano y del Caribe de Plantas Medicinales y Aromáticas, 15*(6), 398-406.

Sriti Eljazi, J., Bachrouch, O., Salem, N., Msaada, K., Aouini, J., Hammami, M., Boushih, E., Abderraba, M., Limam, F., & Mediouni Ben Jemaa, J. (2017). Chemical composition and insecticidal activity of essential oil from coriander fruit against Tribolium castaenum, Sitophilus oryzae, and Lasioderma serricorne. *International Journal of Food Properties*, *20*(sup3), S2833-S2845.

Teneva, D., Denkova, Z., Denkova, R., Atanasova, T., Nenov, N., & Merdzhanov, P. (2015). Antimicrobial activity of essential oils and extracts from black pepper, cumin, coriander and cardamom against some pathogenic and saprophytic microorganisms. *Bulgarian Journal of Veterinary Medicine*, *18*(4), 373-377.

Unlukara, A., Beyzi, E., Ipek, A., & Gurbuz, B. (2016). Effects of different water applications on yield and oil contents of autumn sown coriander (Coriandrum sativum L.). *Turkish Journal of Field Crops*, *21*(2), 200-209.

Zamindar, N., Sadrarhami, M., & Doudi, M. (2016). Antifungal activity of coriander (Coriandrum sativum L.) essential oil in tomato sauce. *Journal of Food Measurement and Characterization*, *10*(3), 589-594.

Zeb, A. (2016). Coriander (Coriandrum sativum) oils. In *Essential Oils in Food Preservation, Flavor and Safety*. Academic Press, pp. 359-364.

Zeković, Z., Kaplan, M., Pavlić, B., Olgun, E. O., Vladić, J., Canlı, O., & Vidović, S. (2016). Chemical characterization of polyphenols and volatile fraction of coriander (Coriandrum sativum L.) extracts obtained by subcritical water extraction. *Industrial Crops and Products*, *87*, 54-63.

ABOUT THE EDITOR

Dr. Deepak Kumar Semwal
Assistant Professor
Department of Phytochemistry, Faculty of Biomedical Sciences,
Uttarakhand Ayurved University, Dehradun, India
Email: dr_dks.1983@yahoo.co.in.

Dr. Semwal is currently serving as a faculty member at the Department Phytochemistry, Uttarakhand Ayurved University, India. In addition to the teaching and research, he has been assigned various administrative responsibilities at the University. Prior to joining this position, Dr Semwal has been a Postdoctoral Research Fellow at the Department of Pharmaceutical Sciences, Tshwane University of Technology, Pretoria, South Africa. His research was funded by Tshwane University of Technology and National Research Foundation, South Africa. He received his MSc in Organic Chemistry and PhD in Phytochemistry from HNB Garhwal University, Srinagar, India. During the PhD, he worked on the chemistry and pharmacology of selected plants of Central Himalaya having antidiabetic and antitubercular properties. In 2005, he appointed as Lecturer at the Department of Chemistry, HNB Garhwal University, India. He also worked as a R&D scientist in an India based pharmaceutical industry and his key roles were method developments and technology transfer for various herbal products. In 2010, he received prestigious Dr DS Kothari Postdoctoral Fellowship from University Grants Commission, New Delhi, India. He did his postdoctoral work at the Department of Chemistry, Panjab University, Chandigarh, and his research was based on the investigation of antidiabetic principles from selected plants species from Indian origin. He has been awarded the postdoctoral researcher of the year 2014 by Tshwane University of Technology, South Africa. He has been a PI of seven R&D projects including three ongoing ones funded by different national and international agencies. Presently, he is serving as an editor of many prestigious journals including Journal of Ethnopharmacology (Elsevier) and Current Clinical Pharmacology (Bentham Science). In addition, he has been reviewing for many international journals published by the world renowned publishers. In past, he organized many international and national conferences/ seminars on traditional medicine. He has been published about 100 papers in various journals of international repute and also presented his work at many national and international conferences including as an invited speaker. He has already published 7 books on traditional medicine and phytochemistry. He received many prestigious awards including young scientist for his research and academic performance. His current research interests are phytochemistry, phytomedicine, clinical pharmacology, drug discovery and development, drug delivery and Ayurveda.

INDEX

#

1,2-dimethylhydrazine, xiii, 18

A

abiotic stresses, 96, 103
Acinetobacter baumannii, 22, 79, 87, 88
aflatoxin, 7, 27, 77, 131
agriculture sectors, 115
alkaloids, 10, 16, 24, 27, 115, 119, 195
alliaceae, 113
amino acid, 14, 84, 115
analgesic, 14, 191, 198, 203, 204, 208, 209, 210
Anethum graveolens, 114
angiospermae, 5
anise, 114, 178
anthelmintic activities, 17
anthocyanin(s), 62, 67, 116, 133, 134, 190, 204
antibacterial activity, 31, 34, 64, 74, 75, 76, 77, 78, 79, 80, 82, 83, 86, 89, 90, 91, 92, 168, 192, 208, 222, 224, 225, 230

anticancer, viii, 15, 18, 39, 69, 72, 85, 90, 114, 134, 135, 184, 196, 197, 198, 209, 211
antifertility, 21, 184
antifungal, viii, 14, 21, 28, 39, 69, 75, 77, 80, 84, 86, 87, 88, 89, 93, 95, 99, 114, 122, 124, 134, 168, 171, 179, 192, 197, 198, 200, 223, 229, 230, 234
antihypertensive, 170, 191
antimicrobial, v, vi, viii, 14, 15, 16, 20, 21, 22, 57, 63, 64, 74, 75, 76, 77, 79, 80, 82, 84, 85, 86, 87, 89, 90, 91, 92, 93, 113, 120, 121, 124, 128, 131, 136, 139, 167, 168, 171, 173, 177, 179, 191, 192, 198, 199, 202, 205, 207, 208, 209, 211, 221, 222, 225, 228, 229, 231, 232, 233, 234
antimutagenic, 7, 10, 13, 27, 64, 66, 85, 99, 139, 167, 170, 191, 209, 210
antioxidant, v, xiii, 2, 10, 14, 15, 16, 19, 22, 23, 28, 29, 30, 32, 33, 34, 35, 37, 39, 47, 48, 57, 58, 59, 62, 63, 64, 66, 67, 68, 69, 71, 72, 73, 85, 87, 88, 90, 92, 93, 95, 99, 105, 114, 122, 124, 125, 126, 128, 129, 131, 135, 139, 149, 167, 168, 171, 174, 179, 180, 181, 191, 194, 195, 197, 198,

200, 201, 202, 205, 208, 209, 210, 211, 225, 226, 228, 230, 231
antiviral, 72, 114
anxiety, vii, viii, 13, 15, 17, 31, 39, 70, 90, 98, 99, 191, 202, 210, 226
anxiolytic activity, 15
aphrodisiac, 2, 14, 168, 184, 208, 210
apiaceae, 2, 4, 5, 61, 63, 64, 77, 86, 88, 95, 98, 113, 139, 140, 184, 201, 207, 208, 227, 228
apiales, 5
application(s), v, 1, 2, 10, 27, 32, 34, 39, 43, 48, 73, 85, 86, 95, 98, 110, 114, 116, 117, 118, 119, 120, 123, 129, 134, 136, 140, 141, 143, 145, 169, 170, 172, 173, 175, 179, 180, 184, 198, 204, 207, 208, 217, 218, 219, 220, 230, 231, 234
arteriosclerosis, 22
artificial neural network, xiii, 140, 141, 158, 170, 175, 176, 177, 179
Aspergillus niger, 21, 75, 77, 192, 222
asteraceae, 113
auxin metabolism, 101

B

Bacillus cereus, 22, 77, 83, 222, 223, 225
behenic alcohol, 212
bile acid synthesis, 25
bilirubin, 24
bioactive compounds, 2, 10, 15, 21, 24, 26, 29, 37, 38, 49, 64, 85, 116, 117, 147, 171, 173, 184, 186, 189, 194, 195, 227
biological activity, 31, 60, 64, 115, 140, 168, 170, 174, 176, 180, 198
biosynthetic pathway, 117
biotechnological methods, 100
biotechnology, 59, 91, 96, 97, 98, 118, 121, 122, 123, 124, 126, 127, 128, 130, 131, 132, 133, 134, 135, 137, 183
blood glucose levels, 19

borneol, vii, 7, 8, 187, 213, 218
botanical fumigant, 114
butylated hydroxyanisole, 21, 23, 114
butylated hydroxytoluene, 114

C

C. tordylium, 98
cadmium, 41, 42, 44, 110
calcium chloride, 104
calcium-alginate coating, 103
callus, 100, 101, 102, 106, 108, 109, 111, 117, 123, 126, 127, 133, 135
calyciflorae, 5
campesterol, 45, 167
camphor, vii, 8, 66, 73, 78, 139, 163, 164, 170, 194, 207, 212, 213, 215, 217, 218, 219, 223, 224
Cananga odorata, 115, 127
cancer(s), 10, 12, 18, 22, 25, 30, 32, 41, 62, 69, 70, 73, 92, 115, 125, 138, 195, 196, 198, 208, 211
candida, 64, 75, 77, 80, 81, 84, 88, 134, 168, 171, 192, 199, 200, 201, 221, 222, 223, 225, 229, 230, 233
Candida albicans, 64, 75, 77, 80, 88, 168, 201, 221, 222, 223, 225, 230, 233
canned soup, 12
capric acid, 75, 166, 190, 213
carbohydrates, 6, 40, 66, 169
carboxylic acid, 26
cardiovascular disease, 12, 41, 45, 115, 209, 211
carminative, vii, 3, 7, 13, 27, 184, 186, 208, 209, 210, 211
carotenoids, 11, 14, 23, 28, 51, 55, 67, 204
carvone, 8, 213
cataract, 12
cefoxitin, 75, 88
cell colonies, 108
cell division, 100, 102, 105, 106, 108, 110

cellular morphogenesis, 100
cellular totipotency, 96, 100, 101, 105
chemical, 90, 93, 120, 124, 126, 128, 134, 135, 137, 141, 164, 171, 172, 174, 177, 179, 199, 202, 203, 231, 232, 233
chemical composition, 90, 93, 120, 124, 126, 128, 134, 135, 137, 141, 164, 171, 172, 174, 177, 179, 199, 202, 203, 231, 232, 233
chemical composition, viii, 40, 48, 58, 99, 125, 139, 140, 144, 166, 167, 170, 173, 178, 208, 215, 218, 231, 232
chemical mutagens, 103
chemical syntheses, 116
chewing gums, 12
chinese parsley, 5, 9, 207
chitosan, 116
chloramphenicol, 79
chocolate, 48, 208, 210
cholesterol, 5, 6, 10, 11, 13, 18, 25, 27, 29, 39, 40, 41, 45, 85, 91, 99, 167, 186, 194, 200
cholesterol-lowering effect, 99
chutney preparation, 12
cilantro, 5, 9, 13, 14, 65, 78, 83, 87, 173, 180, 185, 207, 230
citronellol, vii, 7, 8, 14, 214
citrostadienol, 45
citrus peel, 3, 66
clevenger apparatus, 142
clinical study, 197
cold, viii, 2, 21, 49, 119, 185
colon cancer, 18, 28, 90, 121, 172, 196, 199
commercial antimicrobial drugs, 115
commercial value, viii, 99
composition, vi, viii, 1, 6, 27, 33, 34, 38, 40, 42, 43, 44, 45, 46, 50, 55, 57, 58, 59, 62, 64, 74, 87, 95, 97, 99, 108, 115, 119, 130, 134, 139, 162, 163, 166, 170, 171, 175, 177, 179, 181, 198, 200, 201, 203, 207, 211, 212, 213, 214, 215, 217, 219, 220, 228, 229, 231, 233

confectionery product, 185
conjunctivitis, 13, 14, 15
conventional breeding, 103, 109
convulsion, vii, 13, 98, 99, 191, 210
coriander seed(s), viii, 2, 3, 5, 6, 7, 8, 9, 13, 14, 15, 16, 17, 18, 19, 22, 24, 25, 27, 28, 29, 34, 39, 40, 41, 43, 44, 45, 46, 57, 70, 73, 74, 75, 78, 99, 121, 122, 123, 140, 141, 147, 149, 156, 159, 162, 163, 164, 165, 169, 172, 174, 175, 177, 180, 181, 186, 187, 191, 193, 194, 200, 207, 208, 210, 213, 215, 216, 220, 223, 224, 225, 226, 229
coriandrones, 7
coumaroylquinic acid, 184, 189
creamed cheese, 12
crude oil content, 43
crude oil content, 219
culinary applications, 2, 38
cumenol, 213
cyclooctacosane, 212
cytokinin, 104

D

dedifferentiation, 100, 101, 103, 105, 129
dhania, 5, 11, 185, 198
diabetes mellitus, 22, 193
diarrhoea, 9, 14, 168, 185, 209, 210
dicotyledonae, 5
dietary fibre, 11, 40, 52, 67, 68, 185
digesting enzymes, 105
dihyrocoriandrin, 7
dill, 58, 87, 114, 173
dimercaptosuccinic acid, 70
diploid cross-pollinated crop, 98
diterpenoid, 116, 126
diuretic, viii, 3, 7, 13, 24, 27, 28, 30, 31, 39, 85, 170, 175, 184, 186, 191, 208, 209, 210, 211
DNA, 110

DNA uptake, 110
dodecenal, 8, 21, 67, 83, 166, 187, 190, 212, 215, 216
DPPH scavenging, 69, 149
drying method, 38, 51, 52, 54, 57, 58, 61, 214, 232
dyspeptic complaints, 99, 191

E

ecological conditions, 100, 231
eczema, 10, 14
eicosane, 214
electrofusion, 110
elicitor-receptor interaction, 117
embryogenesis, 96, 97, 101, 102, 103, 105, 106, 108, 111, 118, 119, 121, 122, 123, 125, 126, 127, 130, 131, 133, 134, 136, 137, 138
endangered plants, 96
Enterococcus faecalis, 22, 83
enzyme level, 107
erysipelas, 10, 13
erythema, 20, 197, 204
Escherichia coli, 53, 57, 64, 74, 77, 78, 83, 86, 88, 222, 223, 224, 225, 229, 231, 233
ethnobotany, 96
extraction, viii, xiv, 27, 39, 44, 47, 53, 64, 115, 116, 123, 140, 141, 142, 143, 144, 145, 147, 148, 149, 150, 154, 155, 156, 157, 158, 159, 162, 163, 166, 169, 170, 171, 172, 173, 174, 175, 176, 177, 178, 179, 180, 181, 187, 214, 215, 216, 217, 224, 227, 230, 233,234
eyesight, 10

F

fatty acids, 6, 7, 14, 38, 40, 43, 59, 61, 67, 99, 115, 140, 166, 176, 186, 189, 191, 194, 226, 228

fennel, 73, 90, 114, 126
fertilizers, 115, 217, 218
fever, 3, 10, 11, 14, 21, 185, 210
flavonoids, vii, 7, 10, 14, 15, 16, 24, 27, 32, 34, 37, 99, 116, 149, 176, 188, 189, 190, 193, 195
flavour industry, 95, 99
flow cytometry technique, 111
foeniculum vulgare, 86, 114, 126, 176
foliage, 1, 9, 10, 11, 12, 13, 20, 23, 27, 33, 49, 51, 52, 55, 61, 216, 232
folk medicine, vii, 17, 85, 170
food poisoning, 14, 77, 78, 84, 90
food processing, 114, 172, 177
foods, 9, 11, 23, 24, 25, 28, 34, 39, 48, 49, 50, 52, 53, 55, 60, 61, 64, 66, 73, 86, 88, 91, 120, 121, 172, 178, 185, 204, 208
furan, 7
furosemide, 24

G

gallic acid, xiv, 10, 68, 69, 148
gallic acid equivalent, xiv, 68, 148
gas chromatography, 67, 71, 174, 216
genetic potential, 100
genetic transformation, 96, 110, 111, 127
genotype, 105, 108, 122, 220
Geotrichum, 21
geraniol, vii, 7, 8, 78, 139, 163, 164, 170, 187, 207, 213, 214, 215, 217, 218, 219
geranyl acetate, 7, 8, 67, 164, 193, 207, 213, 215, 217, 218, 225
germination frequency, 104
germplasm, 96, 97, 101, 103, 104, 113, 118, 122
ginsenoside, 116, 135
glazonoids, 7
glucose, 19, 106, 107, 169, 174, 193, 201
Glycyrrhiza inflata, 116, 137
glycyrrhizin, 116, 137

gram-negative, 22, 64, 75, 77, 78, 79, 83, 84, 168, 192, 222
gram-negative bacteria, 78, 83, 84, 168, 192

H

haemonchus contortus, 16, 29, 193, 197, 200
headaches, viii, 13
heavy metals, 14, 41, 116
Helicobacter pylori, 79, 90
heneicos-1-ene, 20, 23, 33
heneicosanol, 212
hepatoprotective, 7, 15, 24, 25, 27, 32, 33, 39, 67, 169, 177
herbals, 64
herbicides, 114
herbivores, 116
heterocyclic compounds, 7
high-quality product, 100
Hippocrates, 2, 13
HPLC, 55, 69
human body, 64, 195
human dermal fibroblasts, 71
hydrocarbon, 113, 164
hydrodistillation, viii, 76, 140, 141, 142, 145, 147, 162, 179, 187, 215, 216, 231, 233
hyperlipidemia, 26, 44, 186, 194
hypoglycemic, 7, 13, 14, 27, 172, 193, 198, 199, 205, 208

I

in vitro, 96
in vitro culture, 96, 98, 100, 103, 112, 115, 117, 124, 134, 135, 190
in vitro experimental model, 100
in vitro tissue culture, 96
indomethacin, 22
infectious diseases, 115

inflammation, viii, 10, 19, 23, 168, 186
insomnia, vii, viii, 10, 13, 15, 98, 99, 191, 210
insulin, 15, 18, 19, 29, 124, 169, 174, 175, 193, 200
International Code of Botanical Nomenclature, 3, 30
Iranian coriander, 217
iron, 6, 10, 11, 14, 15, 41, 42, 44, 51, 66, 67, 68, 71, 92
isocoumacinvizcoriandrin, 7
isoprene, 113, 163

J

joint pains, 184, 186

K

kaempferol, 10, 184, 188, 189, 190
Kasbour, 207
ketoconazole, 192
Kinza, 5
Klebsiella pneumonia, 64, 75, 77, 78, 83, 88, 192

L

Lactobacillus, 224
lamiaceae, 113, 128
Lasioderma serricorne, 227, 234
laxative, viii, 209, 211
limonene, vii, 7, 8, 73, 139, 163, 164, 170, 207, 213, 214, 215, 217, 218, 225
linalool, vii, 6, 7, 8, 15, 22, 23, 39, 66, 69, 78, 83, 85, 88, 92, 95, 99, 139, 162, 163, 164, 166, 168, 170, 187, 191, 192, 193, 194, 196, 199, 204, 207, 212, 213, 214, 215, 216, 217, 218, 219, 220, 221, 222, 223, 224, 225, 227, 228, 230

linnaeus, 4
lipid peroxidation, 71, 72, 114
lipophilic, 45, 62, 113
lipopolysaccharide, xiv, 20
loss of appetite, vii, 10, 13, 98, 99, 210
luteolin, 10, 188, 189
lymene, 7

M

Maceration, 141, 143
magnesium, 5, 6, 10, 11, 41, 42, 44, 51, 62, 67
malonic acid, 113
mannitol, 106, 107
measles, 185
medicinal activities, 99
medicinal plants, 23, 31, 90, 92, 96, 97, 115, 121, 126, 131, 195, 205
Melastomam alabathricum, 116
membrane permeability, 113, 192
metabolic pathways, 117
methyl group, 113
methyl jasmonate, 116, 126, 133, 134, 135, 137
methyl-D-erythritol-4-phosphate, 113
methylmercury, 26
meticillin-resistant, 22
mevalonic acid, 113
microbial cells, 116
microbiota, 53
microcarpum, 3, 4, 98, 216
micropropagation, 97, 120
microwave drying, 50, 54, 55, 61
mineral content, 50
minimum bactericidal concentration, xiv, 21, 80, 83
minimum inhibitory concentration, xiv, 74, 79, 81, 84, 222
moisture content, 21, 52, 55
molecular biology, 112, 117

Monilia sitophila, 21
monoterpene hydrocarbons, 7, 8, 99
monoterpenes, vii, 80, 99, 163, 164, 165
morphogenesis, 100, 103, 122
morphogenetic processes, 111
Mucor, 21
multiple drug/chemical resistance, 115
multiplication, 96, 97, 100, 101, 103
mutations, 111
myrcene, 8, 187, 213, 215, 217, 218
myristicin, 8, 214
Myrtaceae, 113

N

naringin, 188
natural antioxidants, 9, 16, 73, 114, 133, 195, 226
natural products, ix, 115, 125, 175
nausea, 3, 15, 21, 185, 210
neochidilide, 7
nerol, 8, 164, 187, 213, 218
nerolidol, 8
Nerylacetate, 214
neurologic disorders, 23
niacin, 6, 42
nitric oxide, 20, 226
nitrogen, 2, 31, 145, 218, 232
nonanal, 8, 212
nutraceuticals, 2, 34, 61, 91, 178, 180, 204
nutritional profile, 1, 2, 5

O

octadecane, 8
oil, 43
optimization, 62, 98, 107, 108, 140, 141, 147, 148, 149, 155, 158, 159, 160, 161, 162, 169, 171, 172, 177, 178, 179, 180, 181, 233
organogenesis, 96, 97, 127, 129

organoleptic properties, 38, 49, 53
osmotic balance, 113
osmotic compounds, 107
ounshavu, 5
oxygen, xiv, 22, 47, 49, 113, 129

P

pain receptors, 191
palmitic acid, 43, 78, 166, 189
Panax ginseng, 116, 135
pathogenic microorganisms, 115
Penicillium expansum, 21
Penicillium lilacinum, 84, 192
Penicillium stoloniferum, 21
perfumes, viii, 9, 75, 115, 208
persil arabe, 5
pest management, 115
pesticides, 114, 145
petroselinic acid, 6, 38, 39, 43, 78, 85, 166, 213, 216
pharmaceutical industries, 114, 209
pharmaceutical industry, 96, 97, 211, 235
pharmacological properties, 1, 2, 15, 191
phenolics, 69, 99, 113, 149, 208
phenylpropanoid synthesis, 118
phenylpropanoids, 114
phosphatidylglycerol, 186
phospholipid, 18, 46, 186, 197
phosphorus, 5, 41, 42, 44, 67, 68, 218, 231
phthalides, vii, 7, 18
phytomedicine, 87, 115, 134, 176, 236
piles, 13
Pimpinella anisum, 114
pinene, vii, 7, 8, 15, 67, 73, 78, 118, 164, 168, 187, 207, 212, 213, 215, 217, 218, 219, 224
piperaceae, 114
plant genetics, 115
plantaricin CS, 82, 192
plating density, 107, 108

plumb, 41, 42, 44
pluripotentiality, 105
poaceae, 113
polygenic agronomical traits, 110
polyherbal formulation, 186, 204
polypetalae, 5
polyphenols, 10, 15, 16, 23, 27, 32, 184, 187, 188, 189, 191, 202, 203, 234
post-harvest drying, 115
potassium, 5, 6, 10, 11, 41, 42, 44, 51, 67, 219, 230
pressure potential, 106
primary metabolites, 115
proanthocyanidins, 184
processing, v, ix, 37, 38, 48, 56, 57, 58, 60, 62, 63, 86, 180, 199, 230
pro-inflammatory mediator, 20
pro-interleukin, 20
protein, xiv, 6, 20, 40, 49, 50, 55, 56, 67, 68, 70, 71, 117, 171, 210
protein denaturation, 49
protocatechinic acid, 16, 190, 195
protoplast, 105, 106, 107, 108, 109, 110, 113, 118, 119, 120, 121, 123, 126, 129, 130, 132, 136, 137
protoplast division, 108, 137
provitamin, 12
Pseudomonas aeruginosa, 22, 64, 74, 75, 77, 83, 192, 224, 225
pyrazine, 7
pyridine, 7

Q

quercetin, 10, 15, 25, 169, 184, 188, 189, 190, 195
quinic acid, 195

R

radical scavenging, 23, 33, 67, 69, 89, 91, 195, 201, 226, 232
random amplified polymorphic DNA, 112
reactive oxygen, xiv
reactive oxygen species, xiv, 22, 69, 92, 115, 195
resorcinol, 188
rheumatism, vii, 13, 14, 21, 39, 168, 184, 210
Rhizopus stolonifer, 21
riboflavin, 6, 11, 42
Russian coriander, 4, 213
rutaceae, 113

S

Saccharomyces cerevisiae, 222
salicylic acid, 116, 190, 204
Salmonella, 20, 22, 64, 77, 78, 83, 223, 225
Salmonella typhi, 20, 22, 64, 77, 78, 83, 223, 225
Salmonella typhimurium, 22, 83, 223, 225
Salvia sclarea, 116, 126
Sanskrit literature, 2
saturated fat, 40, 43
saturated fatty acids, 40, 43, 61, 166
secondary metabolites, 16, 96, 97, 98, 113, 115, 117, 128, 129, 132, 134, 137, 140, 184, 190, 191
senescence, 117
sexual incompatibility, 96, 109
shelf life, 38, 48, 56, 62, 70, 115
Sitophilus oryzae, 194, 227, 228, 229, 233
sleeping disorder, 17
smallpox, 14, 185
soaps, 9
sodium, 5, 6, 11, 42, 44, 51, 67, 70, 91, 104
sodium alginate, 104
soft drinks, 208, 210

somatic embryos, 101, 103, 104, 109, 119
sorbitol, 106, 107, 223
source tissue, 107
Soxhlet extraction, 140, 142, 143, 147, 217
Spice Board of India, 3
staphylococcus aureus, 22, 64, 74, 75, 77, 78, 83, 86, 88, 192, 222, 223, 224, 225, 229
sterols, 7, 18, 25, 45, 59, 140, 166, 167, 186, 187, 194
stigmasterol, 45, 167, 187, 213
stomach disorders, 3, 9, 21
storage conditions, 73, 107
storage time, 104
Streptococcus infection, 13
Streptococcus pneumoniae, 22
streptozotocin, 19, 29, 128, 174, 200
substrate, 80, 107, 221
sucrose, 107, 133
sugars, 40, 67, 106, 115
sulfur, 42, 44
Swiss albino mice, 17
synergistic effect, 70, 88, 192
synthetic seed gel matrix, 104
synthetic seeds, 103, 104, 118

T

tartaric acid, 10, 189
terpenes, viii, 113, 163, 166, 208
terpenoids, 113, 117, 142
terpinene, vii, 7, 8, 15, 66, 163, 164, 187, 193, 207, 213, 214, 215, 216, 217, 218, 225
tetradecane, 212
tetrahydrofuran, 7
therapeutic, viii, 11, 13, 114, 196
thiamine, 6
thujene, 8
tissue, 96
tobacco, 9, 11, 12, 209

tocopherol, 42, 46, 73, 187
total fat, 6, 40, 189
total phenol contents, 190
totipotency, 97, 100, 116, 121
toxic compounds, 108, 202
traditional Greek medicine, 2
tranquillizer, 194
trans fatty acids, 40
trans-cinnamic acid, 187
transgenic plants, 101, 111
Tridecanal, 212, 213
turmeric powder, 15

U

ultrasound-assisted extraction, xiv, 144, 178, 180
Umbellale, 5
Umbelliferae, vii, 5, 83, 88, 89, 113, 207, 208
unsaturated fatty acids, 41, 57, 59
urinary tract infections, 15, 66

V

vaccenic acid, 43
vacuum, 52
vacuum drying, 50, 52, 62
Vanillin, 214
venshivu, 5
vicenin, 10, 189
vitamin B, 5, 6, 27, 42, 68
vitamin C, 5, 6, 27, 40, 41, 42, 52, 67, 68
vitamin D, 5, 6, 27, 42
volatile nature, 114
volatile oils, 79, 87, 116

Y

yeast infection, 64, 192

Z

zinc, 5, 6, 42, 44, 67, 110, 120
zygotic embryos, 96, 101, 121, 130

Related Nova Publications

PLANT DORMANCY: MECHANISMS, CAUSES AND EFFECTS

EDITOR: Renato V. Botelho

SERIES: Plant Science Research and Practices

BOOK DESCRIPTION: Dormancy is a mechanism found in several plant species developed through evolution, which allows plants to survive in adverse conditions and ensure their perpetuation.

HARDCOVER ISBN: 978-1-53615-380-4
RETAIL PRICE: $160

MICROPROPAGATION: METHODS AND EFFECTS

EDITOR: Valdir M. Stefenon, Ph.D.

SERIES: Plant Science Research and Practices

BOOK DESCRIPTION: Plant micropropagation is one of the most classical and widespread biotechnological tools used around the world. Undoubtedly, this technique brought quite important advances to our knowledge about morphological, physiological and developmental patterns of plants, to the progress of genetic breeding and to the establishment of the genetic engineering, among others.

SOFTCOVER ISBN: 978-1-53614-968-5
RETAIL PRICE: $82

To see a complete list of Nova publications, please visit our website at www.novapublishers.com

Related Nova Publications

GERMINATION: TYPES, PROCESS AND EFFECTS

EDITORS: Rosalva Mora-Escobedo, PhD, Cristina Martinez, and Rosalía Reynoso

SERIES: Plant Science Research and Practices

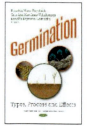

BOOK DESCRIPTION: *Germination: Types, Process and Effects* is a book that brings together the contribution of new and relevant information from many experts in the fields of food and biological sciences, nutrition, and food engineering, to provide the reader with the latest information of fundamental and applied research in the role of edible seeds and discuss the benefits of consuming them.

HARDCOVER ISBN: 978-1-53615-973-8
RETAIL PRICE: $230

PHYTOCHEMICALS: PLANT SOURCES AND POTENTIAL HEALTH BENEFITS

EDITOR: Iman Ryan

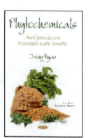

SERIES: Plant Science Research and Practices

BOOK DESCRIPTION: The opening chapter of *Phytochemicals: Plant Sources and Potential Health Benefits* discusses macronutrients and micronutrients from plants along with their benefits to human health.

HARDCOVER ISBN: 978-1-53615-478-8
RETAIL PRICE: $230

To see a complete list of Nova publications, please visit our website at www.novapublishers.com